Lecture Notes in Biomathematics

Edited by S. Levin

2

Mathematical Problems in Biology

Victoria Conference

Edited by Pauline van den Driessche

Springer-Verlag
Berlin · Heidelberg · New York 1974

Dr. Pauline van den Driessche
Department of Mathematics
University of Victoria
Victoria, British Columbia
Canada
V8W 2Y2

Library of Congress Cataloging in Publication Data
Main entry under title:

Mathematical problems in biology.

(Lecture notes in biomathematics, v. 2)
Bibliography: p.
1. Biomathematics--Congresses. I. Van Den Driessche, P., 1941- ed. II. Title. III. Series.
[DNLM: 1. Biology--Congresses. 2. Mathematics--Congresses. 3. Models, Biological--Congresses. W1LE334 v. 2 / QA37 M432 1973]
QH323.5.M37 574'.01'51 74-13574

AMS Subject Classifications (1970): 92 XX, 92A02, 92A05, 92A15

ISBN 3-540-06847-3 Springer-Verlag Berlin · Heidelberg · New York
ISBN 0-387-06847-3 Springer-Verlag New York · Heidelberg · Berlin

Printed in Germany.

Offsetdruck: Julius Beltz, Hemsbach/Bergstr.

FOREWORD

A conference on "Some Mathematical Problems in Biology" was held at the University of Victoria, Victoria, B.C., Canada, from May 7 - 10, 1973. The participants and invited speakers were mathematicians interested in problems of a biological nature, and scientists actively engaged in developing mathematical models in biological fields. One aim of the conference was to attempt to assess what the recent rapid growth of mathematical interaction with the biosciences has accomplished and may accomplish in the near future. The conference also aimed to expose the problems of communication between mathematicians and biological scientists, and in doing so to stimulate the interchange of ideas. It was recognised that the topic spans an enormous breadth, and little attempt was made to balance the very diverse areas.

Widespread active interest was shown in the conference, and just over one hundred people registered. The varied departments and institutions across North America from which the participants came made it both academically and geographically mixed. The chief activity of the conference was the presentation of papers. Nine invited guest speakers (see table of contents) each gave a one hour talk. These covered a wide range of topics. There were twenty-five shorter (twenty minute) contributed papers, and almost all papers were followed by a five minute question and discussion period. Duplicated abstracts of presented papers were available at the meeting. An evening informal discussion meeting of participants, chaired by Dr. A. B. Tayler, and led by Drs. E. M. Hagmeier, E. C. Pielou and V. Klee, was also scheduled, and proved to be a very worthwhile session. There were several happy social occasions, and many unscheduled group and individual discussions.

It is naturally not possible to reproduce all these discussions, but most of the papers which were presented are included here. Out of the total number of papers presented at the meeting, over half appear in full in this volume. Others appear as summary articles, and a few appear as short abstracts which were submitted for the meeting. Where presented papers appear in full elsewhere, the references have been included. Nearly all the papers printed were typed at the institution of the respective speakers, and were subsequently edited only. Dr. P. Rajagopal's article on the relevance of mathematics to research and teaching in the biological sciences was stimulated by the evening discussion meeting. Opinion at this meeting also brought out the need for a selected bibliography of this subject, and such a bibliography, compiled by Dr. G. W. Swan, is included in this volume. Also included are some problems posed at the meeting.

It is hoped that this volume will serve both those who attended the conference and those unable to attend, and that this permanent record will add to the stimulation and dialogue generated at the meeting. In particular it is intended that it will acquaint the reader with various mathematical ideas and techniques which have been used in problems of a biological nature, and introduce some unsolved problems arising in this field. The number of pages devoted to a particular area in no way reflects the importance of the area, but rather the interest of participants.

Thanks are due to the National Research Council, the Leon and Thea Koerner Foundation, and the University of Victoria for their financial support which made the conference possible. The University of Victoria provided facilities for the conference, and many people of the University community gave freely of their help and advice. Many thanks to them, to the conference participants, and all who made the meeting such a success. In addition, thanks to the conference speakers for their cooperation in submitting manuscripts, and to the publishers Springer-Verlag.

P. van den Driessche
University of Victoria
Victoria, British Columbia

TABLE OF CONTENTS

An asterisk (*) indicates the author who presented the paper

MEASUREMENT PROCESSES AND HIERARCHICAL ORGANIZATION*

L.M. Bianchi, Graduate Faculty of Mathematics and

Atkinson College, York University, Toronto, Ontario.

Abstract

One of the most difficult problems encountered in the study of complex structures, in particular of biological systems, is that of the origin of the apparent arbitrariness of controls and constraints in the multi-level description of such systems. An examination of some typical examples leads to the suggestion that whenever such arbitrariness is present, decision processes are essentially responsible for it. Since measuring processes are intimately related to decision processes, it is proposed that hierarchical organization is a consequence of the mutual interaction of several measuring systems simultaneously engaged in some form of observation.

1. SYSTEMS, COMPLEXITY, AND ARBITRARINESS

The very concept of complex systems is unfortunately still lacking any precise definition. Intuitively we consider biological entities, social structures, economic processes, behavioral patterns, natural languages, etc. as typical instances of complexity, whereas atoms, crystals, glaciers or even galaxies appear, at least in comparison to the former, to be relatively simple systems. If we accept such a distinction, we may indeed suggest that complex systems are essentially biological, if we allow such a word to encompass collective properties of living systems, including symbolic behavior. This convention has the advantage that we can delay confronting the question of the thresholds beyond which complexity, as it were, emerges.

In order to endow our definition with more substance, let us consider first a physical system, for example the scattering of two billiard balls. A dynamical description in this case amounts to an explicit knowledge of the potential energy of the

*This work was supported by Grant A6747 from the National Research Council of Canada.

two particles as a function of their relative positions and momenta, as well as of the laws of motion (Newton's equations). The only arbitrariness refers to the choice _we_ have in setting up the experiment, in preparing the system. It is important to notice that such choice is completely independent of the dynamical laws (8). But what is even more important is that the essential behavior of the system is conversely completely independent of the initial conditions.+ Let us now consider a dilute gas in equilibrium inside a container. We are familiar with two possible ways of describing its behavior - particle physics and thermodynamics. Within the framework of the former, the gas is pictured as a collection of particles freely interacting among themselves and with the particles composing the walls of the container. Each particle obeys the usual equations of motion with a suitable potential function. Of course it is impossible, even in principle, to specify the initial conditions of all the particles, let alone to solve all the equations of motion. In order to gain any understanding of the behavior of the system we must renounce detailed information, introduce probabilities and perform averages (statistical mechanics). A new form of arbitrariness arises here, for we must operate a selection among the infinite averages possible, and there is no criterion intrinsic in particle or statistical mechanics to dictate or even to suggest a recipe for a choice. As it is well known, thermodynamics provides such an instrument and points to pressure, temperature, volume, etc. Notice that, historically, thermodynamics developed much earlier. The fact that _we_, the observers, are equipped with a perceptual apparatus which lends _immediate_ (i.e. not mediated) operational meaning to these variables, is obviously responsible for such historical sequence. I will return to this point later on. Let us finally consider a third system - a busy street intersection with a traffic-actuated traffic light. In this case the dynamical laws of motion of the cars are _not_ independent of the initial and boundary conditions. More precisely, the laws of motion (not just the kinematics) of individual cars are constrained by an arbitrary

---

+ The system is dynamically invariant with respect to those transformations which represent changes in the initial conditions.

function of the motion itself of all cars. More importantly, such an arbitrary function is generated indogenously by the system (by the car drivers) and is physically and stably realized in the system (the traffic light) (3,6). It is indeed this property which contributes to the fundamental distinction between the billiard balls and the gas systems on the one hand, and the street intersection system on the other. In this sense the latter is said to be complex (i.e. biological, according to our definition).

2. DECISION SYSTEMS

Before turning my attention to actual biological systems (in the more traditional sense of the word biological) I wish to consider briefly what is meant by a decision process. I will do so by examining two illustrations of the range which such a concept spans. We can describe the propagation of an electromagnetic wave through a medium simply on the basis of Maxwell's equations and the properties of the medium. We can also obtain the actual path of the wave through a variational technique (the least action principle). Such a technique essentially amounts to choosing among all possible paths the one which minimizes the total action. In other words, we assign to each path a quantity (action) and we decide for which path such a quantity is lowest. It is clear, however, that in this case the decision process is but another form of the laws of motion, and thus no real decision insofar as the laws of motion do not offer any substantial alternatives. Consider instead the brachistochrone problem: to determine that path along which a material point, under the action of gravity, moves from a given point to another lower given point in the shortest possible time. In this case it is the boundary conditions (not the law of motion) which are chosen so as to minimize a certain quantity (time). From the point of view of the law of motion such optimization is arbitrary (in fact it is a question mathematicians ask, not the system). In this case we speak of a proper decision process. So are the choices we underscored in the illustrations of the previous section.

The question I wish now to pose is this: is there evidence that decision systems are physically realized in actual cases of biological interest? Can we find biological systems in which arbitrariness is not due to the particular epistemological approach of the observer, but is generated by the system itself? More generally and fundamentally, are there instances of biological systems which are descriptions of other related biological systems (6)? In order to begin to explore such a query we must first consider the hierarchical nature of biological (more generally, complex) systems. Clearly the sequence: molecule - macromolecule - organelle - cell - tissue - organ - organism - population - etc., is a hierarchy in the usual sense of the word. We can choose to describe a biological system at any of these levels and, if we believe that such an observational and descriptive choice is entirely ours, the only task left is to establish ways of relating such descriptions to each other. To put it in other words, the existence and nature of the levels is entirely arbitrary from the point of view of the system and depends exclusively on the observer. My question thus enquires whether biological hierarchies are, in fact, intrinsic in biological systems, whether the observer is, in fact, built in such systems.* The test of such a hypothesis consists in finding a biological system which (i) exhibits arbitrariness, i.e. behavior which cannot be explained in terms of its function at a certain level, and (ii) is engaged simultaneously in other functions defining other levels, such that within the framework of the latter the explanation of arbitrariness is provided. I will touch here only briefly on a couple of examples, while a fuller analysis will appear elsewhere(1). Let us consider any enzyme from the point of view of both its amino acid sequence and its folded conformation. It is well known that the former does not uniquely determine the latter. More important, however, is the fact that the function of the folded structure is extremely specific. From the point of view of protein synthesis this property is arbitrary, since it is a consequence of residual degrees of freedom of the bonds between amino acids which

---

*In this sense I am here considering the problem of subjective vs. objective description.

are utilized by the protein's environment. Such arbitrariness leads to a simplification (5,6) of the enormous complication of the enzyme molecule and to the utilization of this simpler property as such in the regulation of biochemical reactions. Thus the initial arbitrariness receives explanation not at the physico-chemical level (which is essentially local) but at in least in part at the next one where averages of non-local properties takes place (in part, of course, at the evolutionary level which, it is important to remember, acts on the phenotype not on the genotype). Another example is provided by the cells of the potato tuber whose proliferation is controlled by a particular gene which produces iso-amylalcohol which in turn inhibits cell division through a rate-dependent threshold mechanism (4). Such inhibition depends only on the concentration of a volatile alcohol about the tuber cells not on the details of molecular motions. The very concept of concentration is arbitrary from the point of view of individual chemical event, while not from the point of view of the cell population level.

3. OBSERVATION AND DECISIONS

Let us return to our examination of decision processes. From the examples we have considered I will try to generalize this notion in the following way. Suppose we have a system S which can exist in a set of states $\{S_i\}$. Given a state S_i I will indicate by R_{ij} the set of all possible evolutions from S_i. Such relationships must be compatible with the law of motion which describes the dynamics of the system when supplemented by initial and boundary conditions or constraints. Notice that some of these constraints may be non-holonomic, in which case the dynamics of the system is itself a function of such conditions (6,7). If now T is a mapping of the $\{R_{ij}\}$ onto (this is the notion of simplification) a set $\{V_k\}$ and A is an algorithm which specifies a rule for selecting one particular member V_k^* of such set, and thus through T^{-1} one particular relationship R_{ij}^*, we will say that the quintuplet $(\{S_i\}, \{R_{ij}\}, T, \{V_k\}, A)$ represents a decision process. If not only S but also the other components are physically realized we will speak of a decision system.

Now it can be shown that decision processes and measurement and observation processes are intimately related (7). The connection is essentially established by noticing that measurement means classification and that classification is a form of decision making. In a forthcoming paper I will formalize this notion and relate it to the above notation(1). What emerges is that if a (sub)system S is simultaneously observed by two other (sub)systems M and N which are in turn interacting between themselves, the evolution of S will be determined not only by its law of motion but by other constraints which will appear arbitrary from the point of view of the dynamics of S, but not so from the more comprehensive point of view of S+M+N. The mappings of the type T and the algorithms of the type A will appear as physical properties of the various interactions between S and M, S and N and M and N. Some of these mappings will, for example, involve some kind of ensemble averages on a collection of systems like S and thus enter essentially in the dynamical law of S (7). The important corollary of these considerations is that we are not confined to only two observing systems M and N, but can keep adding more to the chain, thus originating a hierarchy which is indefinitely continuable (3,6). The linkage between the levels of such a hierarchy is not simply the relationship of inclusion, but an actual form of control on the part of certain levels on the degrees of freedom and the dynamical behavior of other levels. We speak thus more accurately of a _continuable control hierarchy_ (3). It becomes then clear why it is very difficult, if not impossible, to reduce the explanation of the behavior of a particular subsystem (or of the whole system, for that matter) to the language appropriate to a definite level of complexity (e.g., biochemistry), unless, of course, we introduce _ad hoc_ phenomenological considerations. These, however, will always appear arbitrary - the only cure consists in recognizing that the subsystem is simultaneously participating in the dynamics of a different level (2).

References

1. Bianchi, L. M. In Proceedings of the Conference on the Mathematics & Physics of the Central Nervous System. Trieste 1973. In press.

2. Bianchi, L. M., and Hamann, J. R. Math. Biosci. 5, 277 (1969).

3. Bianchi, L.M., and Hamann, J. R. J. Theor. Biol. 28, 498 (1970).

4. Bonner, J. In Pattee, H., ed. Hierarchy Theory. New York: G. Braziller (1973).

5. Levins, R.: ibidem.

6. Pattee, H. H. In Rosen, R., ed. Textbook of Mathematical Biology. New York: Academic Press (1971).

7. Rosen, R. Math. Biosci. 14, 151 (1972).

8. Wigner, E. P. In The Nobel Prize Lectures. New York: American Elsevier (1964).

ANALYSIS OF WAVE PROPAGATION IN CILIA AND FLAGELLA

J. J. Blum and J. Lubliner

Department of Physiology and Pharmacology, Duke University Medical Center, Durham, North Carolina 27710 and Department of Civil Engineering, University of California, Berkeley, California 94720

A. Introduction

The purpose of this symposium, I take it, is to bring to the attention of applied mathematicians problems from biology which might present challenges of intrinsic interest to the applied mathematician and which necessitate a rather sophisticated treatment if they are to be fully understood. One such example is the analysis of wave motion in flagella. In the jargon of applied mathematics, this is a problem in wave propagation in a non-linear system. Perhaps the most important point for the mathematician to be reminded of is the importance of acquiring a good grasp of the physiology of the system before formulating models. Here I shall merely give a very brief summary of current thinking on the nature of the mechanochemical system responsible for flagellar movement. The subject has recently been reviewed at length (Blum and Lubliner, 1973) and many of the ideas alluded to in the present paper are more fully developed and documented therein.

Figure 1 shows a cross section of the almost universal structure of the flagellum, the 9x2 axonemal complex. Although many variants on this basic structure are known, the 9x2 configuration is noteworthy for its stability through the evolutionary history of the eukaryotes, i.e., from protozoa to man. Since motile sperm are known with 9x0 and 9x1 configurations, the central pair of singlet microtubules appear not to be essential. The evolutionary advantage that presumably inheres in the 9x2 configuration is not yet apparent.

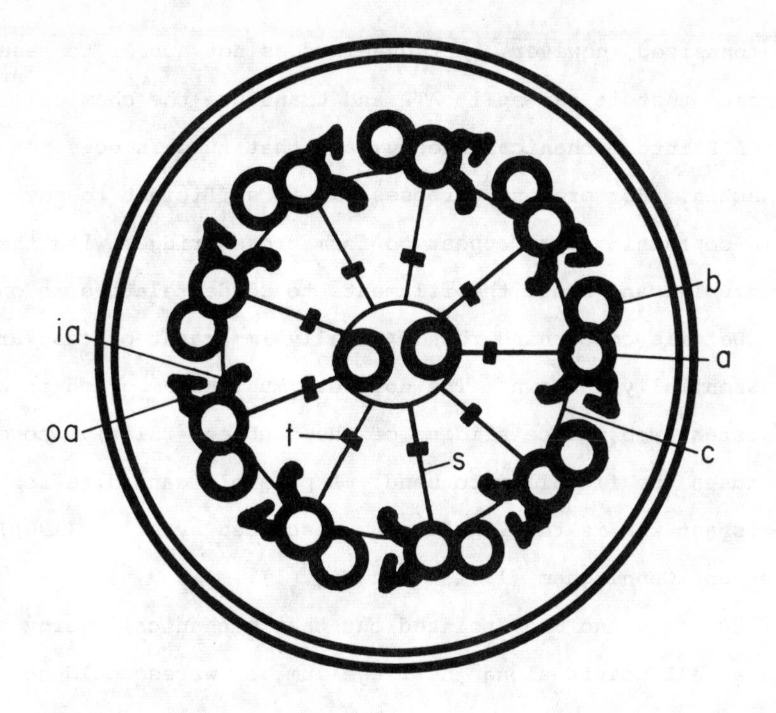

Figure 1. Cross section of a cilium of Tetrahymena pyriformis: a and b are subfibers a and b, respectively, of the outer doublets; c, the connection between the outer doublet fibers (nexein); oa, the outer arm; ia, the inner arm; s, radial spoke; t, thickened region along spoke. Taken from Allen, 1968.

The basic unit of both the singlet microtubules and the outer 9 doublet microtubules is a dimer composed of 2 similar but not identical proteins, called tubulins, each about 55,000 daltons molecular weight. The A-subfiber of each outer doublet bears two non-identical arms spaced vertically about 170 Å apart. These arms point towards the B-subfiber of the adjacent doublet. It is thought that the major protein component of these arms is the mechanochemically active ATPase, dynein. Recent evidence (Summers and Gibbons, 1971) shows that the basis of flagellar motility is a sliding filament mechanism, in which the axonemal doublets and dynein appear to play roles analogous to actin and myosin, respectively, in skeletal muscle. It must be emphasized, however, that dynein does not appear to resemble myosin except that it can split ATP and transduce the chemical free energy of ATP into mechanical energy, and that tubulin does not resemble actin. For present purposes, it is sufficient to say that the dynein-containing arms appear to form cross bridges with the adjacent doublet and cause the filaments to slide relative to one another. Details concerning this crucially important process are at present essentially unknown. The doublets must be embedded in a shear-resistant web, since sliding of the doublets relative to one another causes the flagellum to bend. A possible candidate for the shear-resistant web is the link system described by Allen (1968) and by Dentler and Cunningham (1971) (see Fig. 1).

Over 20 years ago Grey pointed out that mechanical energy must be fed in at all points along the flagellum, or waves could not be maintained with non-decreasing, or, in some cases, increasing amplitudes. This was put on a quantitative basis by Machin (1958), who showed that unless energy was fed in locally, wave amplitude would diminish markedly within about 1 wavelength, contrary to observation. It was originally assumed that flagella beat with a sin wave motion. Brokaw and Wright, however, showed that the outlines of sperm tails on high speed photomicrographs were not sinusoidal, but consist of straight segments alternating with regions of uniform curvature (Brokaw and Wright, 1963). The transition between a bent and unbent region is very sharp. This deviation from sinusoidal shape makes little difference for computations of swimming speed or power output (reviewed in Blum and Lubliner, 1973) but, as Brokaw (1965) pointed out, implied a new concept for the way in which bending was propagated. Ahead of a region in which active bending is occurring the flagellum is deformed passively. Maximum passive bending will occur in the region immediately adjacent to the actively bent region. If the passively bent region "fires" (i.e., begins active bending) when a critical radius of curvature is attained, a self-propagating wave will move down the flagellum. The system is comparable to nerve impulse in the sense that once a critical threshold of disturbance is achieved, a non-linear event occurs which releases energy into the system locally and thus drives the adjacent region to threshold. Brokaw (1966a) derived an equation of motion for a simplified flagellum consisting of two fibrils connected by shear-resistant cross bridges and a contractile element. Lubliner and Blum (1971a) extended this analysis by allowing a more realistic geometry, by utilizing a contractile system that could be due to a local contraction mechanism or a sliding filament mechanism, and by including the effects of viscosity in the formulation. A slightly condensed derivation of the equation of motion in an infinite flagellum in a viscous medium follows.

B. The Equation of Motion for an Infinite Flagellum

Consider a flagellar model consisting of n longitudinal elements or fibers connected by shear-resistant cross bridges; the flagellum bends in a plane, the local radius of curvature being $\rho=1/\mu$. Let u be the local coordinate in the plane of bending (see Fig. 2) perpendicular to the axis, measured from the centroid of all the contractile elements and positive away from the center of curvature. Let ε_i denote the longitudinal strain at the centroid of the ith element and u_i its location; then the strain at any point of the element is

$$\varepsilon = \varepsilon_i + \mu(u-u_i). \tag{1}$$

If the axial force in the element is F_i and its cross sectional area is A_i, the axial stress is

$$\sigma = \frac{F_i}{A_i} + E\mu(u-u_i) \tag{2}$$

where E is Young's modulus for the contractile material. Recent measurements suggest that E is about $5\text{-}9\times10^{10}$ dyne/cm^2 (Baba, 1972); this rather high value may refer to the properties of the active force

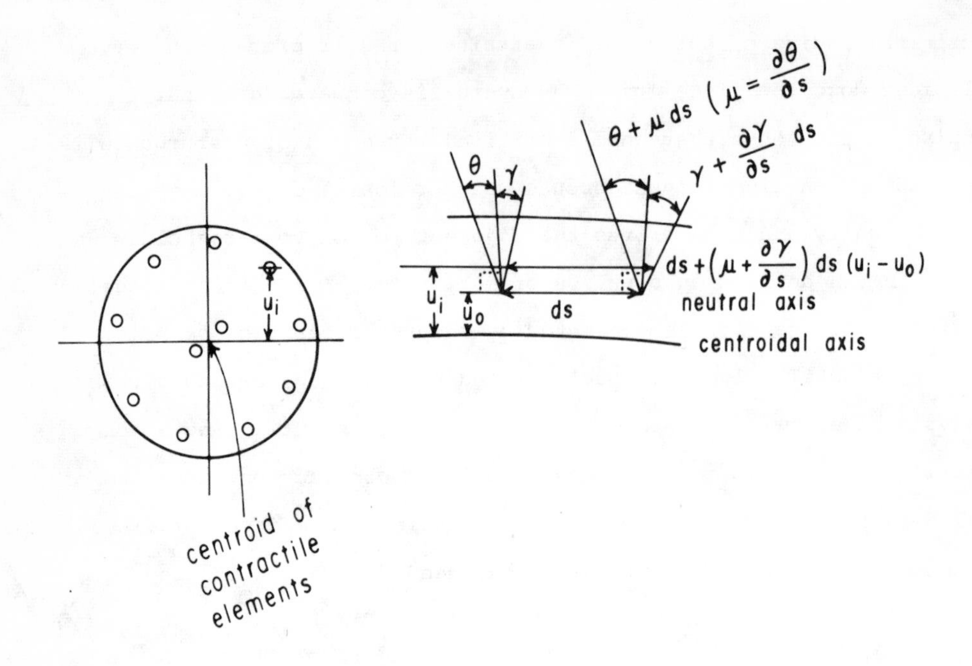

Figure 2. Axial strain in a flagellar model. The left side of this figure shows a schematic cross section of the flagellum; u_i is the distance from a line perpendicular to the axis and passing through the centroid of the ith element. The right side of the figure shows the disposition of the elements in the plane of the bend. For further details, see text.

generating elements rather than the elastic structures in parallel with the active elements, as discussed by Brokaw (1972). The resultant bending moment is

$$M_i = \sum_{i=1}^{n} F_i u_i + E\mu \sum_{i=1}^{n} I_i \tag{3}$$

where I_i is the moment of inertia of the ith element. The effective shear force exerted by the cross bridge is

$$S = \frac{\partial}{\partial s} \sum_{i=1}^{n} F_i u_i \tag{4}$$

The shear angle γ is considered to be made up of an elastic part, S/GA_s (where G is the shear modulus of the cross bridge and A_s is its effective cross sectional area) and a part due to active sliding, γ_s.

$$\gamma = \gamma_s + S/GA_s = \gamma_s + \frac{1}{GA_s}\frac{\partial}{\partial s}\sum_{i=1}^{n} F_i u_i \qquad (5)$$

After eliminating the strain ε and imposing the equilibrium conditions $\sum_{i=1}^{n} F_i = o$ and $M_i + M_v = o$, where M_i and M_v are the internal and viscous moments, respectively, one obtains the following equation for the curvature μ.

$$\frac{2\alpha_o}{h^2\alpha_2}\frac{\partial^2\mu}{\partial s^2} = (1 + \frac{2\alpha_o}{\alpha_1})\mu - \frac{c}{h} + \frac{M_v}{\alpha_1} - \frac{1}{h^2\alpha_2}\frac{\partial^2 M_v}{\partial s^2}, \qquad (6)$$

where

$$2\alpha_o = E\sum_{i=1}^{n} I_i \qquad (6a)$$

$$\alpha_1 = E\sum_{i=1}^{n} A_i u_i^2 \qquad (6b)$$

$$h = \sum_{i=1}^{n} A_i u_i^2 / A_c u_c \qquad (6c)$$

$$\alpha_2 = GA_s/h^2 \qquad (6d)$$

$$c = c_o - h\,(\partial\gamma_s/\partial s) \qquad (6e)$$

This equation specifies the curvature at any point along the flagellum in terms of rigorously defined elastic and shear constants for any specified geometry of the fibrils in terms of the viscous bending moment and a general contraction process, c. In words, α_o is the sum of the bending resistance of each element expressed in terms of its own moment of inertia; α_1 is the bending resistance of the whole flagellum; h is the effective diameter of the cilium with A_c being the area of the active elements ($m<n$) and u_c being equivalent to the position of the centroid of the active elements; α_2 is the shear resistance of the flagellum; and c is the total effective contraction. For a sliding filament mechanism, c_o, the change in length (per unit length) due to local contraction, is zero. Up to this point the analysis is quite general except for one tacit assumption, i.e., that the bridges undergo an active shear deformation, γ_s, in addition to

a passive elastic shear. An active sliding mechanism could alternatively be based on the assumption that an additional set of bridges (comparable to the actomyosin bridges of muscle), are formed in parallel with the passive bridges during the active process. Lubliner (1973) has shown that this formulation, which is used by Brokaw in his recent computer simulation studies of flagellar motion (Brokaw, 1972), is mathematically equivalent to the assumption that each bridge can undergo an active and a passive shear.

To proceed further one must derive an explicit expression for M_V. If V_T is the longitudinal velocity, V_N the transverse velocity, and θ the angle at any point of the flagellum with respect to a given reference line, then

$$\frac{\partial \theta}{\partial t} = \frac{\partial V_N}{\partial s} + \mu V_T \tag{7}$$

Since $\mu = \partial\theta/\partial s$, equation (7) becomes

$$\frac{\partial \mu}{\partial t} = \frac{\partial^2 V_N}{\partial s^2} + \frac{\partial}{\partial s}(\mu V_T) \tag{8}$$

Lubliner and Blum (1971a) neglected the term in μV_T in this equation, thus limiting the following analysis to waves of small amplitude. Removal of this restriction would be most welcome. The transverse viscous force per unit length can be written

$$W = \kappa\eta V_N = \partial^2 M_V/\partial s^2 \tag{9}$$

where η is the viscosity of the medium in which the flagellum is undulating and κ is a geometrical factor. $\kappa\eta$ is frequently called the transverse drag coefficient and often written as C_N. The geometrical factor κ has been evaluated by hydrodynamiscists (Burgers, 1938).

$$\kappa = \frac{4\pi}{\ln\left(\frac{2L'}{d}\right) + \delta} \tag{10}$$

In equation (10) L' is the length and d the diameter of the flagellum, and δ is a number of order 1 that depends on the shape of the motion. For a flagellum 20 μm long and 0.25 μm in diameter, the $\ln$ term is $\sim$ 5, and one makes little error no matter how one chooses

δ. Since $\delta = -0.5$ for a long thin filament bearing sin waves, the error will be negligible for even a short flagellum undulating in an approximately sinusoidal fashion. On combining equations (8) and (9) one obtains

$$\frac{\partial^4 M_V}{\partial s^4} = \kappa\eta \frac{\partial \mu}{\partial t} \tag{11}$$

which may be used to eliminate M_V from equation (6). Before doing this it is convenient to introduce the steady wave assumption, namely that μ and c, which are functions of s and t, depend only on the combination x=s-vt, where v is the (constant) wave speed. The resulting ordinary differential equation is

$$\frac{2\alpha_o}{h^2\alpha_2}\,\mu^{vi} - \left(1+\frac{2\alpha_o}{\alpha_1}\right)\mu^{iv} - \frac{\kappa\eta v}{h^2\alpha_2}\,\mu^{iii} + \frac{\kappa\eta v}{\alpha_1}\,\mu^{1} = -\frac{1}{h}\,c^{iv}\ , \tag{12}$$

where $\mu^i = d^i\mu/dx^i$.

This is a 6th order equation which, for a given c(x) can be solved by means of a Fourier transform. The solution will then depend on the roots of a quintic equation. A detailed analysis of this equation would be useful. In the absence of such an analysis, and especially since we have already limited the equation to waves of small amplitude, we used Brokaw's (1966a) notion that the viscous bending moment is distributed sinusoidally with a wavelength L, and we shall approximate

$$\frac{d^4 M_V}{dx^4} \simeq -\left(\frac{2\pi}{L}\right)^2 \frac{d^2 M_V}{dx^2} \simeq \left(\frac{2\pi}{L}\right)^4 M_V \tag{13}$$

Instead of equation (12), then, one obtains the much simpler equation

$$\mu - 2\lambda v\mu' - \frac{\mu''}{\alpha^2} = \frac{\varepsilon}{\alpha^2}\,c \tag{14}$$

The constants α, ε, and λ are written out in Lubliner and Blum (1971a) and need not be copied here.

C. Kinetics of the Sliding Filament System

Equation (14) is the differential equation of motion of a long flagellum moving in a medium of arbitrary viscosity, subject, however, to the limitations which have been noted. Notice that it will not yield a wave solution unless c is non-linear. In other words, it is <u>required</u> that something about the contractile system be non-linear if wave propagation is to occur. For a sliding filament model equation (6e) can be written $c = -h(\partial\gamma_s/\partial s)$. On a steady wave, $\partial\gamma_s/\partial s = \dot{\gamma}_s/v$, where $\dot{\gamma}_s \equiv \partial\gamma_s/\partial t$. The necessary non-linearity is conveniently provided by the assumption that active sliding begins when a critical curvature, μ^*, is attained, and terminates τ seconds later. Available evidence suggests that τ is about 15 to 20 milliseconds and does not vary appreciably with viscosity (Lubliner and Blum, 1971a). Other possible choices for the mechanochemical coupling step will be discussed below. With this choice the contractile process can be described as

$$c = \begin{cases} o & t - t^* < o \\ h\dot{\gamma}_s/v & o < t - t^* < \tau \\ o & \tau < t - t^* \end{cases} \tag{15}$$

where t is the time and t^* is the time at which active contraction begins at a given point. If we define the origin of s such that $t^*=o$ when $x=s-vt=o$, then $t^*=s/v$ and $t-t^*=-x/v$; hence $c(t-t^*)=c(-\frac{x}{v})$. Equation (14) can conveniently be solved by use of Fourier transform. One obtains

$$\mu(x) = \frac{\varepsilon}{\alpha^2}\int_o^\infty K(\xi+x)\; c(\frac{\xi}{v})\; d\xi, \tag{16a}$$

$$\text{where } K(x) = \frac{\alpha}{2\sqrt{1+m^2}}\exp(-\alpha\sqrt{1+m^2}\;|x| - \alpha m x) \tag{16b}$$

Here $m=\ell q$, where $\ell=\lambda/\tau$ is a dimensionless viscosity and $q=\alpha v\tau$ is a dimensionless wave speed. The curvature at any point along the filament can be obtained by inserting equation (15) into (16) and performing the integrations. It is both convenient and reasonable to assume that $\dot{\gamma}_s$ is independent of position within the active region. The integration must be performed in two regions:

(a) For $x>o$, i.e., ahead of the transition point, only passive bending occurs, and

$$\mu(x) = \mu(o)\, e^{-px/v\tau} \qquad (17a)$$

where $p=q(m+\sqrt{1+m^2})$ and $\mu(o)$ is a constant depending on $\eta, \ell, q, \dot{\gamma}_s$, and the mechanical properties of the flagellum. Thus passive bending decays exponentially with distance ahead of the transition point.

(b) For $-v\tau<x<o$. The region where active sliding is occurring must be divided into two sub-regions, $\xi<-x$ and $\xi>-x$. Upon performing the integration, one obtains

$$\mu(x) = \frac{\dot{\gamma}_s q}{2v\sqrt{1+m^2}}\left(\frac{1-e^{-wx/v\tau}}{w} + \frac{1-e^{-p(v\tau+x)/v\tau}}{p}\right) \qquad (17b)$$

where $w=q\,(\sqrt{1+m^2}-m)$.

The value of x at which the curvature is maximum, x_m, can be found by differentiating with respect to x in equation (17b) and setting the result equal to zero. When this is done, one finds that $x_m=-pv\tau/(w+p)$. Substituting this value into equation (17b) yields

$$\mu_{max} = (\dot{\gamma}_s/v)\ [1-e^{-q/2(\sqrt{1+m^2})}] \simeq \dot{\gamma}_s/v, \qquad (18)$$

since, for the range of viscosities and wave speeds encountered, the exponential term in equation (18) is small compared to unity. Thus the maximum curvature is directly proportional to the rate of active sliding and inversely proportional to wave speed. For sufficiently large w, equation (17) shows that the curvature is close to μ_{max} over most of the active region, in agreement with the observation that the active region approximates a circular arc (Brokaw and Wright, 1963). Examination of Brokaw's (1966b) data for _Chaetopterus_ sperm shows that the maximum curvature is inversely proportional to wave speed. For _Chaetopterus_ sperm, then, the sliding filament model is consistent with the data if $\dot{\gamma}_s$ is chosen to be a constant. For living _Lytechinus_ and _Ciona_ sperm μ_{max} is essentially independent of viscosity even though v decreases and η increases (Brokaw, 1966b). For these sperm, therefore, it is reasonable to choose $\dot{\gamma}_s$ proportional to wave velocity. A possible generalized expression for $\dot{\gamma}_s$ is

$$|\dot{\gamma}_s| = a + bv \tag{19}$$

where a and b are constants. It must be emphasized that while this choice appears reasonable, it is arbitrary. In principle, the kinetics of the sliding process should be prescribed in terms of the rate of cross bridge cycling and the amount of sliding generated per cycle, both of which may be functions of curvature, rather than the steady wave velocity. Gibbons and Gibbons (1972) have recently shown that extraction of the outer arms of sea urchin sperm flagella reduces frequency towards half while leaving wave shape parameters relatively unaltered. This suggests that beat frequency is largely controlled by the number of arms (i.e., cross bridges), whereas the amount of sliding occurring in each cycle is controlled largely by elastic forces within the axonemal structure. Both waveform and frequency are known to depend on ATP concentrations (see, e.g. Douglas and Holwill, 1972). Indeed at high concentrations of ATP the frequency and wave speed of glycerol-extracted _Lytechinus_ sperm decreases as viscosity increases, while the curvature of the bent region remains approximately constant. As ATP concentration is reduced, however, the curvature of the bent regions begins to _increase_ with increasing viscosity, as observed in living _Chaetopterus_ sperm at high viscosity. This suggests that the variation in waveform seen between different species at low viscosity and the variation seen as a response to increasing viscosity reflect the interrelation between ATP concentration and the passive and active components of the basic motile system common to most, if not all, eukaryotic flagella.

D. The Mechanochemical Coupling Function

The simplest choice for mechanochemical coupling is the assumption used above that active sliding begins when the curvature approaches a critical value, μ^*. This can be called the simple trigger or fixed threshold hypothesis. For _Chaetopterus_ sperm the use of this hypothesis, in conjunction with $\dot{\gamma}_s = a$ leads to satisfactory agreement with the experimental data. For *Ciona* sperm, however, where the choice $\dot{\gamma}_s = bv$ is indicated by the experimental data, one finds that

$$\mu(o) = \frac{b}{2}\left(1 - \frac{m}{\sqrt{1+m^2}}\right) .$$

If $\mu(o)$, which is the curvature at the point of transition (see eqn 17a) is set equal to μ^*, then m would necessarily be a constant. Since $m = \ell q$ this would require that q be inversely proportional to ℓ. The experimental data for the sperm of *Ciona* and *Lytechinus*, however, do not show an inverse proportionality between wave speed and viscosity. If we are to retain the choice $\dot{\gamma}_s = bv$ for *Ciona* and *Lytechinus* sperm, some modification of the mechanochemical coupling is required. In the region immediately ahead of the transition point the sliding filament mechanism is activated. This activation need not be a simple trigger, but instead may depend on the prior history of passive bending. Some experimental support for this notion comes from recent experiments of Brokaw and Gibbons (1972), who found that the beating of different regions of a flagellum reactivated by ADP may not be completely independent. An alternate form of the mechanochemical coupling, which allows for a history-dependent activation process is as follows: instead of the curvature itself, some linear functional over its history (Volterra, 1959) for $t < t^*$ attains a critical value, say $\bar{\mu}$. This can be written as

$$\bar{\mu} = \int_{-\infty}^{t^*} M\,(t^*-t)\,\frac{\partial \mu}{\partial t}\,dt \tag{20}$$

where $M(t^*-t)$ is a function which can be thought of as the memory function for the mechanochemical coupling process. The simplest form of $M(t^*-t)$ is

$$M(t^*-t) = 1 + c(t^*-t) \qquad \text{for } t < t^* \tag{21}$$

where c is a constant. If c=o one obtains from equation (20) the memory-independent triggering of the active process, with $\bar{\mu}$ equivalent to μ^*. If the integration in equation (20) is performed for the case when $c \neq o$, one obtains

$$\bar{\mu} = (1 + \frac{c\tau}{p})\ \alpha\ \dot{\gamma}_s \tau / 2p\sqrt{1+m^2} \tag{22}$$

For any prescribed value of $\bar{\mu}$ and for a particular choice of the kinetics of the active sliding process, such as in equation (19), this equation can be shown (Lubliner and Blum, 1972) to be an implicit relation between the dimensionless wave speed, q, and the dimensionless viscosity, ℓ, and thus may be used to fit experimental data. Further insight into the meaning of the memory function with respect to cross bridge action may be gained by the following considerations. Suppose that the event leading to the triggering of active contraction is the attainment of a critical magnitude, ε^*, of the strain in a cross bridge. The strain, ε, depends on the corresponding stress, which in turn may be proportional to the curvature μ. If the stress-strain relation is that of a Maxwell material (modeled by a linear spring and a linear dashpot in series), then the stress-strain relation is given by

$$\dot{\varepsilon} = a_1 \dot{\mu} + a_2 \mu \tag{23}$$

Hence

$$\varepsilon(t) = a_1 \mu(t) + a_2 \int_{-\infty}^{t} \mu(t')\ dt' = \int_{-\infty}^{t} [a_1 + a_2(t-t')]\ \frac{\partial \mu}{\partial t'}\ dt' \tag{24}$$

and

$$\varepsilon(t^*) = \varepsilon^* = \int_{-\infty}^{t} [a_1 + a_2(t^*-t)]\ \frac{\partial \mu}{\partial t}\ dt\ . \tag{25}$$

Equation (25) is equivalent to equations (20) and (21) if $c = a_2/a_1$. Thus $1/c$ may be considered as a relaxation time characterizing the stress-strain properties of a cross bridge under active deformation.

Equation (22), as stated above, amounts to an implicit relation between wave speed and viscosity. For the sperm of <u>Chaetopterus</u>, the choice $\dot{\gamma}_s = a$ and $c = o$ leads to the fit shown in Fig. 3 between the experimental data and the theory. From this fit one computes

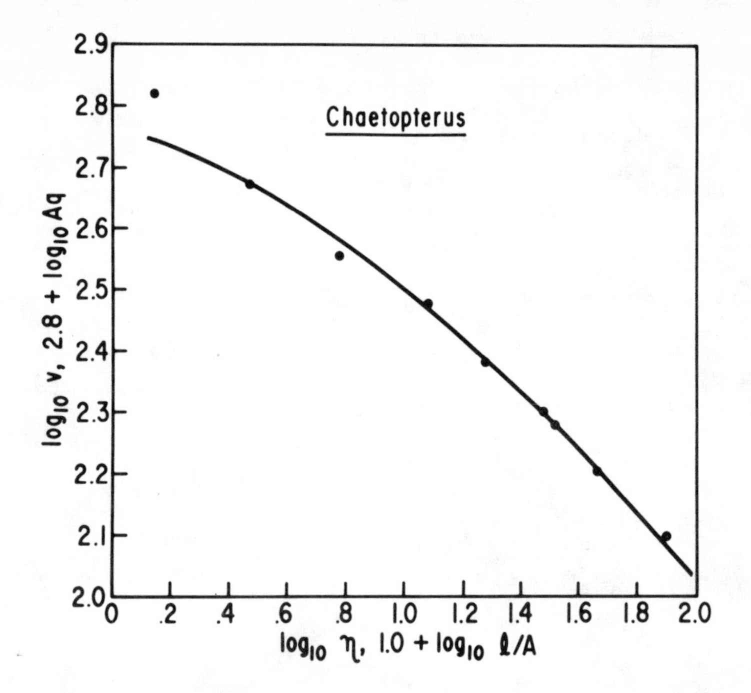

Figure 3. Comparison between experimental and theoretical wave speed viscosity for the sperm of the marine worm Chateopterus variopedatus. the solid points are Brokaw's (1966b) data for wave speed, v, and adjusted viscosity, η. The line is the theoretical curve, with ℓ and q representing the dimensionless viscosity and wave speed, respectively, and A a theoretically derived fitting parameter.

$\dot{\gamma}_s = a = 119\ \sec^{-1}$ for the rate of the active sliding process in this sperm. Other parameters of the sperm tail may also be deduced from the theory, but will not be gone into here. The theory also predicts that for $\dot{\gamma}_s = a$ the radius of curvature of the bent region should be proportional to wave speed; that this proportionality occurs in the experimental data is shown in Table 1. Although the present theory gives a satisfactory fit to the data on curvature of the bent region and on wave speed, it must be emphasized that the theory in its present form makes no statement concerning the frequency of beating. It is

TABLE 1

Effect of Viscosity on Radius of Curvature of Bent Region and on Wave Speed for Sperm of *Chaetopterus*

η centipoise	ρ µm	v µm/sec	ρ/v sec
1.4	4.23	660	.0064
3.0	3.7	470	.0079
6.0	3.2	360	.0089
12.0	2.6	300	.0087
19.0	2.0	240	.0083
28.0	1.6	200	.0080
33.0	1.5	190	.0079
46.0	1.46	160	.0091
80.0	1.23	125	.0098

now known that spontaneous beating may occur in any portion of the flagellum (Brokaw and Gibbons, 1972; Lindemann and Rikmenspoel, 1972a, 1972b). Thus bending on one side must automatically initiate bending on the opposite side of a flagellum. Brokaw (1970, 1972) has set up a computer program for the modeling of flagellar waves in which the active shear moment generated at a point is controlled by the curvature of the flagellum at that point. The model is not restricted to low amplitude motions and includes the effects of both boundary conditions and initial conditions. Such a model generates frequency as well as wave speed and wave form and is a powerful tool for the study of flagellar motility. In so far as it uses parameters (such as an internal viscous bending resistance) which may not have any straightforward connection with structural and biochemical information on cross bridge action, it may be difficult to interpret the meaning of the parameters in molecular terms.

Perhaps the most important step in the analytical approach

exemplified in this present work is to find a form for $\dot{\gamma}_s$ which would activate the contractile mechanism to slide in the opposite direction after an appropriate time delay, thus generating the frequency in terms of parameters which might have relatively straightforward meanings in molecular terms. Once the frequency is specified, knowledge of the wave speed at each viscosity, plus the assumption of a viscosity-independent duration of the contractile process and a fast relaxation is sufficient to largely determine wave length and wave amplitude.

For the sperm of Ciona and Lytechinus the choice $\dot{\gamma}_s$=bv in conjunction with a history-dependent mechanochemical coupling yields a good fit to the wave-speed and wave-form data as a function of viscosity (Figs. 4, 5). The value of c in the memory function

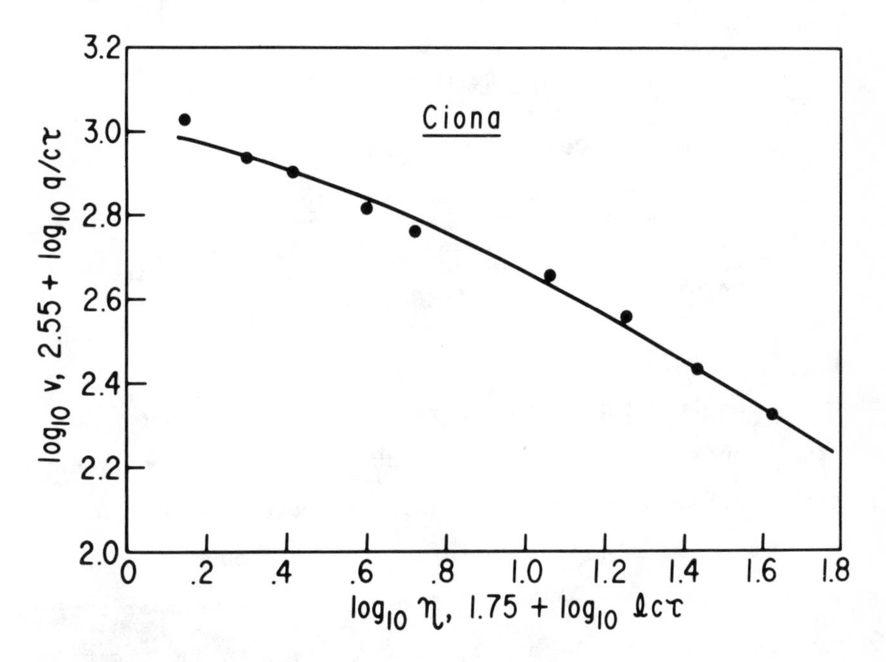

Figure 4. Comparison between experimental and theoretical wave speed-viscosity data for the sperm of the tunicate Ciona intestinalis. The solid points are Brokaw's (1966b) data for wave speed, v, and adjusted viscosity, η. The line is the theoretical curve, with ℓ and q dimensionless viscosity and wave speed, respectively, τ the duration of the active sliding process, and c the relaxation time for the memory function.

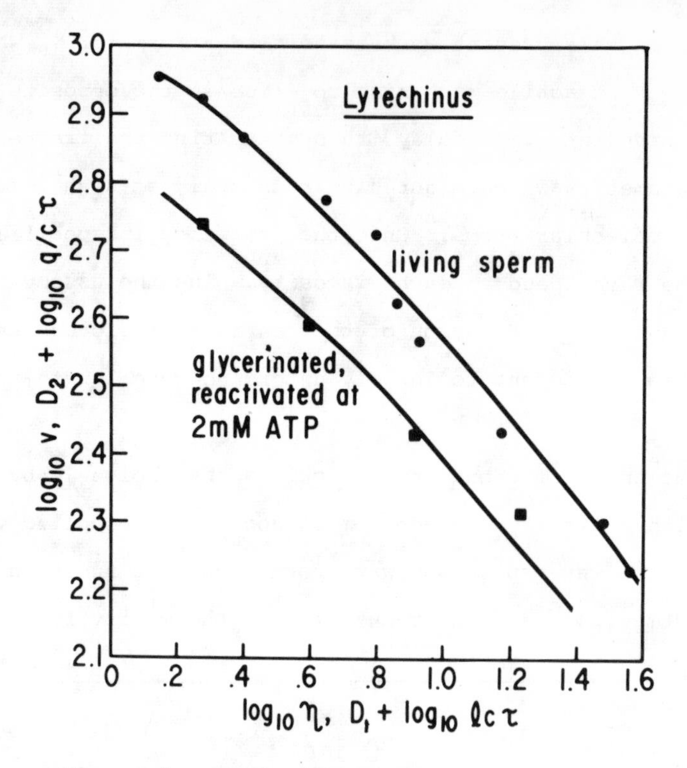

Figure 5. Comparison between experimental and theoretical wave speed-viscosity data for sperm of the sea urchin Lytechinus pictus. Solid symbols are Brokaw's (1966b) data for wave speed and adjusted viscosity; lines are theoretical curves. D_1 and D_2 are fitting parameters. For further details see legend to Figure 4 and Lubliner and Blum (1972).

(equation 21) required to obtain the fit shown was $\sim 10^4$ sec^{-1}. Although this appears to be large, the integrand of the memory integral (equation 20) decays very fast, and in fact the memory integral plays a role for only a brief time before critical curvature is obtained. Because of this the exact form of $M(t)$ is unimportant; it is only required that $M(t)$ be approximated by $1+ct$ in the first few milliseconds.

E. Bend Initiation and Boundary Conditions

We have so far restricted our attention to the propagation of steady waves on an infinite flagellum. While this may be a useful approximation to the situation on long flagella, it is unlikely to be suitable for analysis of the bending in short flagella or cilia. Cilia were originally classified as different from flagella because of their asymmetrical oar-like motion, in contrast to the symmetrical undulation of flagella. It is now known, however, that cilia and flagella have the same ultrastructure and that the same organelle can beat as a cilium or a flagellum (see Blum and Lubliner, 1973 for review). Indeed, segments of sperm tail can be reactivated to beat in a ciliary-like manner or like a flagellum depending on whether the end is constrained or unconstrained (Lindemann and Rikmenspoel, 1972a, b). Thus boundary conditions at the end of a flagellar filament determine the kind of bending cycle. Although bend initiation can occur at any place along a flagellum (Brokaw and Gibbons, 1972; Goldstein, Holwill and Silvester, 1970), it generally occurs at one end which acts as a pacemaker for the entire flagellum. Lubliner and Blum (1971b) have extended the theoretical approach outlined above to a semi-infinite flagellum in a medium of zero viscosity. Neglect of the term in λ in equation (14) yields the following second-order differential equation

$$\frac{d^2\mu}{dz^2} - \alpha^2\mu = -\ \varepsilon\ c(z,t), \tag{26}$$

where we have used z instead of s as the coordinate of length along the flagellum. At the time that this treatment was first done there were no data on bend initiation in cilia or flagella as a function of viscosity. Recent elegant work by Baba and Hiramoto (1970), however, provides data on the curvature along a (compound) cilium during the entire beating cycle, and thus provides a strong incentive for the applied mathematicians to develop an analysis for the semi-infinite domain using equation (14) rather than the low-viscosity approximation equation (26).

The complementary solution to equation (26) is

$$\mu_c(z,t) = A(t)e^{-\alpha z} + B(t)e^{\alpha z} \tag{27}$$

but B(t) must be zero since wave amplitude remains finite. By the method of variation of parameters or by Fourier transform procedures it can be shown that the particular solution to equation (26) is

$$\mu_p(z,t) = \frac{\varepsilon}{2\alpha}\int_0^{z^*(t)} e^{-\alpha|z-\xi|}\, c(t-t^*(\xi))\, d\xi\,. \tag{28}$$

We assume that c depends on z only through the initial condition, c=o when $t>t^*(z)$. Here $t^*(z)$ is the time at which $\mu=\mu^*$ at the location z, and, conversely, $z^*(t)$ is the coordinate of the critical bending point at any instant t. It is convenient to choose z=o at the point of insertion of the flagellum into the cell and t=o as the time when, at z=o, the critical radius of curvature is attained (i.e., $\mu(o,o)=\mu^*$). Self propagation will be assured once the critical radius of curvature is attained.

The general solution to equation (26) is obtained by adding the complementary and particular solutions:

$$\mu(z,t) = A(t)\, e^{-\alpha z} + \frac{\varepsilon}{2\alpha}\int_0^{z^*(t)} e^{-\alpha|z-\xi|}\, c(t-t^*(\xi))\, d\xi\,. \tag{29}$$

Upon introducing the boundary condition

$$\mu(o,t) = \mu_o(t)\,, \tag{30}$$

we obtain an expression for A(t) which may be eliminated from equation (29), resulting in

$$\mu(z,t) = \mu_o(t)e^{-\alpha z} +$$

$$\frac{\varepsilon}{2\alpha}\int_0^{z^*(t)} [e^{-\alpha|z-\xi|} - e^{-\alpha(z+\xi)}]\, c(t-t^*(\xi))\, d\xi\,. \tag{31}$$

Substituting this expression for $\mu(z,t)$ in the integral of equation (20) results in an integral equation for $t^*(z)$ or (equivalently) $z^*(t)$ whose solution depends (a) on the choice of active mechanism as embodied in the form of $c(t-t^*)$, and (b) on the nature of the mechanism at the point of insertion as represented by the form of $\mu_o(t)$. The integral equation was originally (Lubliner and Blum,

1971b) solved on a digital computer using the two-state local contraction model (Lubliner and Blum, 1971a) with a simple trigger and a number of arbitrarily chosen forms of $\mu_o(t)$. It was possible, with an appropriate choice of parameters to fit the data reported by Sleigh (1968) for bend initiation in the flagellum of *Codonosiga* as shown in Figure 3 of Lubliner and Blum (1971b). For cilia or flagella in which bends are initiated more slowly, however, no fit was possible. On the other hand, with the sliding-filament model and memory-dependent trigger described above, it appears possible (Lubliner and Blum, 1974) to reproduce most of the ciliary bend initiation data recorded by Sleigh (1968).

Acknowledgement

Dr. Blum is the recipient of a Research Career Development Award (5 K03 GM02341) from the National Institutes of Health. Part of this work was supported by a grant from the National Science Foundation (BG-36865).

References

Allen, R.D. *J. Cell Biol.* 37, 825-831 (1968)

Baba, S.A. *J. Exp. Biol.* 56, 459-465 (1972)

Baba, S.A., and Hiramoto, Y. *J. Exp. Biol.* 52, 675-690 (1970)

Blum, J.J., and Lubliner, J. *Ann. Rev. Biophys and Bioengr.* 2, 181-219 (1973)

Brokaw, C.J. *J. Exp. Biol.* 43, 155-169 (1965)

Brokaw, C.J. *Nature* 209, 161-163 (1966a)

Brokaw, C.J. *J. Exp. Biol.* 45, 113-139 (1966b)

Brokaw, C.J. *J. Exp. Biol.* 53, 445-464 (1970)

Brokaw, C.J. *J. Mechanochem. Cell Motility* 1, 203-211 (1972)

Brokaw, C.J., and Gibbons, I.R. Abstracts of Papers of the 12th Ann. Meeting of the Am. Soc. Cell Biol., p. 29a (1972)

Brokaw, C.J., and Wright, L. Science 142, 1169-1170 (1963)

Burgers, J.M. Proc. K. ned. Akad. Wet. 16, 113-150 (1938)

Dentler, W., and Cunningham, W.P. Abstracts of Papers of the 11th Ann. Meeting of the Am. Soc. Cell Biol., p. 74 (1971)

Douglas, G.J., and Holwill, M.E.J. J. Mechanochem. Cell Motility 1, 213-223 (1972)

Gibbons, B.H., and Gibbons, I.R. Abstracts of Papers of the 12th Ann. Meeting of the Am. Soc. Cell Biol., p. 84a (1972)

Goldstein, S.F., Holwill, M.E.J., and Silvester, N.R. J. Exp. Biol. 53, 401-409 (1970)

Lindemann, C.B., and Rikmenspoel, R. Science 175, 337-338 (1972a)

Lindemann, C.B., and Rikmenspoel, R. Exptl. Cell Res. 73, 255-259 (1972b)

Lubliner, J. J. Theoret. Biol. In press (1973)

Lubliner, J., and Blum, J.J. J. Theoret. Biol. 31, 1-24 (1971a)

Lubliner, J., and Blum, J.J. J. Mechanochem. Cell Motility 1, 15-22 (1971b)

Lubliner, J., and Blum, J.J. J. Mechanochem. Cell Motility 1, 157-167 (1972)

Lubliner, J., and Blum, J.J. To be published (1974)

Machin, K.E. J. Exp. Biol. 35, 796-806 (1958)

Sleigh, M.A. Symposia Soc. Exptl. Biol., Number 22, Aspects of Cell Motility, Academic Press, Inc., N.Y., p. 131-150 (1968)

Summers, K.E., and Gibbons, I.R. Proc. Nat. Acad. Sci. U.S.A. 68, 3092-3096 (1971)

Volterra, V. Theory of Functionals and of Integral and Intergo-Differential Equations. Dover Publications, N.Y. (1959)

MATHEMATICAL BIOECONOMICS

Colin W. Clark

Department of Mathematics
The University of British Columbia

Renewable resource management would seem to provide a fertile area for the application of economic analysis, whether classical or modern. The last of Adam Smith's three sources of wealth - capital, labor and land - for example, clearly includes the class of renewable resources. The fundamental concept of "economic rent" applies directly to the contribution of biological productivity to wealth.

In view of these observations it is surprising how little attention appears to have been given explicitly to the economics of conservation. In fisheries economics, for example, the basic article, published in 1954 (Gordon, 1954), fails to discuss important time-dependent effects. Economic analyses of other industries based on natural animal populations appear to be all but nonexistent. Although forestry and agriculture have perhaps fared somewhat better than average, few people appear to understand the basic economic principles of these subjects.

A possible explanation for this lack of apparent interest in resource economics may lie simply in the "interdisciplinary" nature of the problem. Certainly most of my acquaintances in biology and in economics are quick to admit their ignorance of each other's fields. For this reason it may not be too presumptuous for someone ignorant of both fields, but experienced in mathematics, to try to construct _simple_ mathematical bio-economic models. Such are the subject of this lecture.

Before getting down to specifics, however, I wish to delineate generally the types of models I intend to discuss. They are deterministic models, and they are also dynamic models. Beyond that, the models have two components, one economic and the other biological.

Economically speaking the models are based on the notion of profit maximization, that is, maximization of the present value of net revenues. They are "partial" models in the sense that economic parameters such as price, cost of labor and capital, and so on, are taken to be exogenous.

Biologically speaking, the models are characterized by the models of population dynamics that they incorporate. Three such models will be discussed: the logistic model, a seasonal (discrete time) model, and finally a more complex fishery model utilizing an age-class structure. All three are single-species models in the sense that the population-regulating mechanism is not specified. We might say that the biological analysis, like the economic analysis, is a "partial" analysis, with exogenously determined biological parameters.

The Logistic Model (4)

Let us begin with the simplest and best-known model in population dynamics:

$$\frac{dx}{dt} = f(x); \quad x(0) = x_o. \tag{1}$$

Here $x = x(t)$ represents the size of some chosen population at time t; it may represent either the total number of organisms constituting the population, or its total biomass. The function $f(x)$, the net rate of growth of the population, is assumed to be continuous, concave, and to satisfy

$$f(x) > 0 \text{ for } 0 < x < \bar{x}; \quad f(0) = f(\bar{x}) = 0.$$

(These assumptions are for the sake of simplicity and definiteness; they can be relaxed considerably.) Obviously $\bar{x}$ is a stable equilibrium value for solutions $x(t)$ of Equation (1): it is called the "natural equilibrium population."

Next we introduce a "harvest rate" $h(t) \geq 0$, replacing Equation (1) by

$$\frac{dx}{dt} = f(x) - h(t); \quad x(0) = x_o. \tag{2}$$

If, for example, $h(t) = h = \text{const.}$, then this equation will possess two equilibria, $\underline{x} < \bar{x}$, provided $h < \max f(x)$. The first will be unstable, the second stable. If $h > \max f(x)$ there are no equilibria.

To introduce the economic component of our model, let us assume a constant price $p > 0$ per unit of harvested population, and also a variable-cost function $C(x)$ representing the cost of a unit harvest as a function of the population size x. It will be assumed that $C(x)$ is nonincreasing, i.e. that costs of harvesting can only increase as the population becomes smaller. We also suppose $p > C(\bar{x})$, for otherwise the resource is without economic value. Finally, let $\delta > 0$ denote a given (instantaneous) discount rate.

The total present value of revenues resulting from the harvest rate function $h(t)$ is given by

$$P(h) = \int_0^\infty e^{-\delta t}\{p - C(x(t))\}h(t)dt. \tag{3}$$

The problem we wish to consider is that of determining the harvest rate $h(t)$ that maximizes this expression, subject to the differential equation (2), and also subject to the conditions

$$x(t) \geq 0 \tag{4}$$

$$0 \leq h(t) \leq M; \tag{5}$$

here M is a constant (the "market capacity" for the given resource), although the more general and interesting case $M = M(x)$ is also worth considering.

This is a problem in the calculus of variations. Let us naively use the Euler approach, first substituting $h(t) = f(x) - \dot{x}$ and then using the Euler equation $\partial I/\partial x = d/dt(\partial I/\partial \dot{x})$. The result is the following equation:

$$p - C(x) = \frac{1}{\delta}\frac{d}{dx}\{f(x)\cdot(p - C(x))\}. \tag{6}$$

This equation, which does not involve t, will have a discrete set of solutions $x = \hat{x}_1, \hat{x}_2, \ldots$, each of which represents an extremal equilibrium solution to our variational problem. (For simplicity we shall suppose (6) has at most one solution.) No information about the optimal transition path to the equilibrium can be obtained from Equation (6), however. For this we may use the Pontrjagin maximum principle (Pontrjagin, 1964) as follows.

Introduce the following Hamiltonian function H depending upon a multiplier $\psi(t)$:

$$H(x(t), h(t); \psi(t)) = e^{-\delta t}\{p - C(x(t))\}h(t) + \psi(t)\{f(x(t)) - h(t)\}$$

$$= h(t)\{e^{-\delta t}(p - C(x)) - \psi(t)\} + \psi(t)f(x). \tag{7}$$

Then the following conditions are necessary in order that $h(t) = \hat{h}(t)$ be optimal:

$$\frac{\partial H}{\partial x} = - \frac{d\psi}{dt}; \tag{8}$$

$$H(x(t), \hat{h}(t); \psi(t)) = \max_{0 \leq h \leq M} H(x(t), h; \psi(t)) \text{ for all } t \geq 0. \tag{9}$$

Since (as a consequence of the linearity of our problem) H is linear in the control variable $h(t)$, it immediately follows from (9) that

$$\hat{h}(t) = 0 \text{ or } M \text{ unless } e^{-\delta t}\{p - C(x)\} - \psi(t) = 0. \tag{10}$$

If we assume $e^{-\delta t}\{p - C(x)\} - \psi(t) \equiv 0$ over an open interval (this is called the case of "singular control"),(8) immediately implies that (6) holds for the interval in question.

The final result is as follows.

Theorem 1. Assume that Equation (6) has a unique solution $x = \hat{x}$. Then

$$\hat{h}(t) = \begin{cases} M & \text{whenever } x(t) > \hat{x} \\ f(\hat{x}) & \text{whenever } x(t) = \hat{x} \\ 0 & \text{whenever } x(t) < \hat{x}. \end{cases} \tag{11}$$

A completely elementary, non-variational proof of Theorem 1 is given in the Appendix. The implications of the Theorem are sufficiently obvious: optimal harvesting consists simply of reaching the optimal equilibrium population $\hat{x}$ at the earliest possible instant, and thereafter maintaining a constant population. If Equation (6) has no solution, by the way, then either $\hat{h}(t) \equiv 0$ or $\hat{h}(t) \equiv M$. In the latter case if M is large enough, then $x(t)$ will become zero at some time t, and thereafter perforce we must have $\hat{h}(t) = 0$. This case will be discussed further below.

Let us now return, however, to the basic Equation (6) for the optimal equilibrium population $\hat{x}$. An economist would wish to interpret it as a "marginality condition," and to do this he would have to consider x itself as the basic decision variable. So let us suppose that the population, for the time being, is in equilibrium at the level x. This implies a constant harvest rate $h(t) \equiv f(x)$, and consequently a constant rate of income (or "rent")

$$R(x) = f(x)\{p - C(x)\}. \tag{12}$$

Equation (6) can be written as

$$p - C(x) = \frac{1}{\delta} R'(x). \tag{13}$$

Next the economist will imagine a "marginal" change in the decision variable x, for example a unit decrease to x-1. This will have two effects: it will increase the immediate revenue and it will alter the future rent. The increase in current revenue is simply

$$p - C(x),$$

whereas the change in rent is

$$R(x) - R(x-1) \approx R'(x).$$

The change in rent, however, is to be discounted over all time, resulting in a change in present value equal to

$$\int_0^\infty e^{-\delta t} R'(x)\,dt = \frac{1}{\delta} R'(x).$$

The meaning of Equation (13) is now clear: at the optimal population level $\hat{x}$ the marginal change in current revenue must match the marginal change in the present value of future revenue.

In his book "Resource Conservation: Economics and Policies," the Berkeley economist S.V. Ciriacy-Wantrup (1963) says (p.97):

"Among economic forces affecting conservation, interest [rates] and related forces are among the most powerful, most consistent, and from the standpoint of theoretical analysis and practical effects, among the most clear-cut."

Let us ask how the discount rate* δ affects the solution $\hat{x}$ of Equation (13). First if $\delta = 0$ then we must have

$$R'(\hat{x}) = 0. \tag{14}$$

Optimal harvesting at a zero discount rate thus results in the maximization of rent. It is clear that (14) must always have a positive solution $\hat{x}$.

As $\delta \to \infty$ on the other hand, we will have approximately

$$p - C(\hat{x}) = 0. \tag{15}$$

This equation may not have a solution, but if it does, we reach the conclusion that a high rate of discount results in an equilibrium population at which the rent vanishes (approximately). The strength of the desire for immediate profits results in running the resource down to a level where long-term profits vanish, and this situation is maintained in equilibrium. The long-term rent may be said to have been "dissipated" as a result of the high discount rate.

The process of dissipation of rent has usually been attributed by economists (e.g. Gordon, 1954) to the effects of competition in the exploitation of common-property resources. The present analysis suggests that private resource exploitation with a sufficiently high rate of discount could have a similar effect.

A particularly interesting case arises when (15) has no solution, which will be the case if and only if

$$p > C(0).$$

In this case it turns out that a sufficiently high <u>finite</u> rate of discount δ will prevent Equation(13) from possessing any solution $\hat{x}$, and will therefore imply that "optimal" harvesting results in the extermination of the resource stock. To see this we merely rewrite (13) in the form

---

* The discount rate δ and the "effective rate of interest" i for a unit time interval are related by the formula $\delta = \ln(1 + i)$.

$$\delta - f'(x) = \frac{-C'(x)f(x)}{p - C(x)} = \phi(x). \tag{16}$$

Thus $p > C(x)$ for all x implies that $\phi(x)$ is bounded on $0 \leq x \leq \bar{x}$. Hence (16) has no solution for large δ. More precisely, the following result is easily proved.

Theorem 2. If $\delta < f'(0)$ then (16) has a solution $\hat{x} > 0$. Conversely if $p > C(0)$ and if $\delta > 2f'(0)$ then (16) has no solution.

Notice that $f'(0)$ is the maximum reproductive potential of the population. Theorem 2 thus indicates that species with low reproductive potential are likely to be particularly subject to overexploitation, including possibly extinction. An application to the Antarctic whaling industry is given in Clark (1973c).

A Seasonal Model

Few, if any, natural populations can be modeled realistically by means of the logistic model. Fisheries biologists, for example, usually use seasonal models of population dynamics based on the notion of a "stock-recruitment relationship"

$$x_{k+1} = F(x_k) \tag{17}$$

where x_k denotes the population level at some specific time of the year, such as just prior to spawning. This equation, in which $F(x)$, the stock-recruitment relation, is assumed continuous and increasing in x, is a discrete analog to the logistic equation (1). The economic analysis is quite similar to the previous case, with dynamic programming replacing optimal control theory as the basic mathematical technique. I shall discuss the problem only briefly.

Let $\{h_k\}$ denote a sequence of annual harvests, and modify Equation (17) as follows:

$$x_{k+1} = F(x_k - h_k). \tag{18}$$

The harvests are assumed to satisfy

$$0 \leq h_k \leq x_k,$$

although we could impose an additional limitation $h_k \leq M$ as before. Assume a fixed unit price $p > 0$ and a given cost $C(x)$ for a unit harvest from a population of size x. Then the cost of a harvest h_k from an initial population x_k is approximately

$$\int_{x_k - h_k}^{x_k} C(x)dx = G(x_k) - G(x_k - h_k), \tag{19}$$

where $G' = C$. The present value of the harvest sequence $\{h_k\}$ is thus

$$P(h) = \sum_1^{\infty} \alpha^{k-1}\{ph_k - (G(x_k) - G(x_k - h_k))\} \tag{20}$$

where α is a constant discount factor, $0 < \alpha < 1$.

The optimal harvest policy $\hat{h} = \{\hat{h}_k\}$ can now be described. The basic equation for the optimal <u>post-harvest</u> population $x = \hat{x}$ is

$$p - C(x) = \alpha F'(x)\{p - C(F(x))\}, \tag{21}$$

and this has a similar marginal interpretation to Equation (6). If x_1 is the initial population, and if (21) has a unique solution $\hat{x}$, then the first term of the optimal harvest policy $\hat{h}$ is given by

$$\hat{h}_1 = \max\ (0, x_1 - \hat{x}). \tag{22}$$

This is all the information required to deduce the entire sequence $\{\hat{h}_k\}$, since we have $\hat{x}_2 = F(x_1 - \hat{h}_1)$, and the sequence $\{\hat{h}_2, \hat{h}_3, \ldots\}$ must obviously be an optimal policy with initial population $x = \hat{x}_2$.

As in the continuous logistic model, the optimal harvesting policy causes the population to reach its optimal equilibrium level $\hat{x}$ as rapidly as possible. There is also an analogue to Theorem 2: if $p > C(0)$ and if $\alpha^{-1} > (F'(0))^2$, then $\hat{h}_1 = x_1$ and the population is exterminated at the first opportunity. Elementary proofs of these propositions are given in Clark (1973a).

There is one interesting point of divergence between the seasonal and the logistic models; it concerns the question of rent dissipation under infinite discounting ($\alpha = 0$). By (21) this implies that $p - C(\hat{x}) = 0$ exactly as before. Notice, however, that the annual rent, or net profit, is given by

$$R(\hat{x}) = p\{F(\hat{x}) - \hat{x}\} - \{G(F(\hat{x})) - G(\hat{x})\}.$$

From our assumptions on F and G it is easy to see that this expression is positive for all $\hat{x} \geq x_o$ where $p = C(x_o)$. Thus the rent is never completely dissipated in the seasonal case.

The explanation for this is that, in the seasonal case (unlike the continuous case), profit is generated during the harvest phase, and harvesting only <u>ceases</u> when the population reaches the level x_o at which unit cost equals unit price. By the next harvest time, the population has increased to $f(x_o)$.

The Beverton-Holt Model of a Fishery

In 1957, the British biologists Beverton and Holt (1957) presented a model specifically designed for practical use in the regulation of North Sea demersal fisheries. An economic analysis of this model, which is essentially different from the logistic and seasonal models discussed above, has only been carried out very recently (Clark, Edwards and Friedlaender, 1973).

The Beverton-Holt model assumes first a fixed annual recruitment R of young adult fish, independent of the parent spawning population, at least within the normal range of fishing. This assumption, which is fairly well justified in North Sea fisheries by both theoretical and statistical studies, implies that optimization (or the lack of it) is not associated with the reproductive characteristics of the population. Instead it is the <u>growth</u> and <u>mortality</u> of the fish that is of fundamental importance.

Once recruitment has occurred, the population x(t) is assumed to undergo a constant natural mortality rate $M > 0$:

$$\frac{dx}{dt} = -Mx, \quad x(0) = R. \tag{23}$$

If fishing occurs, it is assumed to add to natural mortality:

$$\frac{dx}{dt} = -(M + F(t))x; \quad x(0) = R. \tag{24}$$

The function $F(t)$ is a measure of fishing intensity, and is taken as the control variable in our analysis.

Individual fish are assumed to grow larger with time; let $w(t)$ denote the (average) weight of a fish at age t years greater than the age $(t = 0)$ of recruitment. For a given year-class of fish, the total "biomass" is thus

$$B(t) = x(t)w(t). \tag{25}$$

We assume that $w(t)$ is a bounded increasing function, and that $w'(t)/w(t)$ is decreasing. The latter condition holds for any concave function, and also for the standard von Bertalanffy growth curve $w(t) = a(1 - e^{-kt})^3$.

It follows that without fishing the biomass $B(t)$ will have a maximum value at some time $t_o \geq 0$. The maximum yield from this single year-class would thus be obtained if the entire population were to be harvested at the instant $t = t_o$. The impractical nature of this solution is obvious.

Let $F(t)$ be a given fishing-intensity function. In the time interval from t to t+dt the yield in biomass is equal to

$$F(t)B(t)dt.$$

Assume a fixed unit price p and suppose that costs are proportional to fishing intensity. Then the net revenue for the time interval (t, t+dt) equals

$$F(t)\{pB(t) - C\}dt.$$

The present-value expression is therefore

$$P(F) = \int_0^{\infty} e^{-\delta t} F(t)\{pB(t) - C\}dt. \tag{26}$$

Assuming that $F(t)$ is restricted by the inequalities

$$0 \leq F(t) \leq F_{max}, \tag{27}$$

we again obtain a linear optimal control problem. As in the case of the logistic model, the optimal solution $\hat{F}(t)$ is characterized by a combined "bang-bang" and equilibrium rule.

To describe the optimal equilibrium we introduce the "net biovalue" function

$$V(t) = pB(t) - C, \tag{28}$$

which represents the net revenue generated (at time t and with population x(t)) by a unit fishing effort, $F(t)dt = 1$. Next we denote by

$$V^{\#}(t) = D_{+}V(t)$$

the right-hand derivative of V(t) under the assumption that $F(\tau) = 0$ for $\tau > t$. Thus $V^{\#}(t)$ is the potential rate of increase of biovalue if no fishing occurs. We then obtain the following characterization of $\hat{F}(t)$.

Theorem 3. Assume that $V^{\#}(t) - \delta V(t)$ possesses at most one change of sign. Then

$$\hat{F}(t) = \begin{cases} 0 & \text{whenever } V^{\#}(t) > \delta V(t) \\ F_{max} & \text{whenever } V^{\#}(t) < \delta V(t) \end{cases}$$

and $\hat{F}(t)$ is chosen to maintain the equality $V^{\#}(t) = \delta V(t)$ whenever this is possible.

An elementary proof appears in Clark, Edwards and Friedlaender (1973).

Bioeconomic equilibrium is thus determined by the simple equation

$$\frac{V^{\#}(t)}{V(t)} = \delta, \tag{29}$$

which equates the rates of relative increase of (potential) biovalue V(t) and relative increase of the value of money, δ. (An analogous formula to this is known in forestry, where V(t) denotes the value of a forest of age t, and Equation (29) determines the optimal age of cutting.)

To see how Theorem 3 applies, we introduce the function

$$\psi(t) = \frac{p^{-1}C\delta}{M+\delta - G(t)} \qquad \text{for } t > t_{\delta} \tag{30}$$

where $G(t) = w'(t)/w(t)$ is the relative growth rate of the fish, and where

$$G(t_{\delta}) = M + \delta. \tag{31}$$

An easy calculation then shows that

$$V^{\#}(t) = \delta V(t) \iff B(t) = \psi(t). \tag{32}$$

The optimal control $\hat{F}(t)$ is best understood by considering the resulting biomass curve $\hat{B}(t)$, and this is indicated in the accompanying sketch. Here $B(t)$ is the

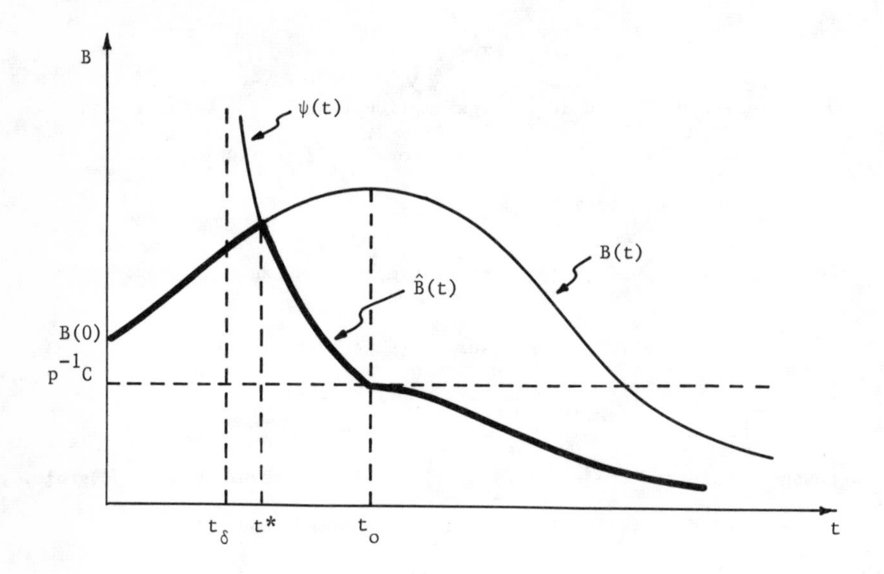

natural biomass curve, having a peak at $t = t_o$. The curve $\psi(t)$ is decreasing, and has a vertical asymptote at $t = t_\delta$. The curves $B(t)$ and $\psi(t)$ intersect at t^*. The optimal control $\hat{F}(t)$ is such that

$$\hat{F}(t) \begin{cases} = 0 \text{ for } t < t^* \\ \text{is positive and decreasing* for } t^* \le t \le t_o \\ = 0 \text{ for } t > t_o. \end{cases}$$

We notice in particular that the "fishing season" must be completed by the time the biomass reaches its natural maximum - this is a consequence of the time-discounting effect. If $\delta = 0$, by the way, then $t_\delta = t^* = t_o$ and all fishing occurs at the single instant $t = t_o$. (This would require $F_{max} = +\infty$, however. Otherwise we would have $\hat{F}(t) = F_{max}$ over some time interval containing t_o.)

---

* This assumes that $w'(t)$ is decreasing.

We also note that as $\delta \to +\infty$, $\psi(t) \to p^{-1}C$, which is the level at which $V(t) = 0$. Infinite discounting thus leads to a fishing "policy" in which the rent is continuously dissipated. In more precise terms we have

$$\lim_{\delta \to \infty} P(\hat{F}) = 0 .$$

In (6) are worked out some figures for the North Sea plaice population. This is a demersal species with a life span of at least 16 years. The age of maximum biomass is $t_o = 13.35$ years. If costs of fishing are relatively low, then the age t^* at which fishing begins is close to the age t_δ of Equation (31). Some values for t_δ are as follows.

δ	0%	5%	10%	15%	20%	25%
T_δ	13.35 years	10.43 years	8.51 years	7.68 years	6.21 years	5.45 years

The value of the discount rate thus has a quite marked effect on the optimal fishing policy, in agreement with the remark of Ciriacy-Wantrup quoted earlier.

So far we have been concerned only with the problem of harvesting a single year-class of fish, whereas in practice most fish populations consist of a mixture of many year-classes. Although some partial results on optimization for the latter problem have been obtained, most of the interesting work remains to be done.

Conclusion

Recent years have seen a significant increase in the understanding of the interdependence of human and "natural" phenomena. In this lecture I have tried to indicate the use of simple mathematical models to illustrate the close interplay between economic and biological forces in problems of resource management. Both biologists and economists have now begun to study such problems seriously. Many of the mathematical tools they will need have already been produced by us mathematicians; in fact I see quite a few of them lying around as "space surplus." I suggest we keep them in good shape, and train plenty of young mathematicians to use them.

Appendix: Proof of Theorem 1.

To change notation slightly, let $\hat{h}(t)$ now denote the control defined by (11), and let $h(t)$ be any other admissible control. We will show that for some $T \leq +\infty$ we have

$$\int_0^T e^{-\delta t} h(t) \{p - C(x_h(t))\} \, dt < \int_0^T e^{-\delta t} \hat{h}(t) \{p - C(x_{\hat{h}}(t))\} \, dt. \tag{33}$$

It then follows from the principle of optimality that h is not optimal.

Suppose first that $x(0) > \hat{x}$, so that $\hat{h}(t) = M$ for small t. Since $h(t) \not\equiv \hat{h}(t)$ we have also $x_h(t) \not\equiv x_{\hat{h}}(t)$; let us suppose for example that

$$x_h(t) \geq x_{\hat{h}}(t) \quad (0 \leq t \leq T), \quad x_h(T) = x_{\hat{h}}(T);$$

$$x_h(t) > x_{\hat{h}}(t) \quad \text{for some} \quad t \in (0,T);$$

the case $T = +\infty$ is included as a possibility.

It follows from the hypotheses that

$$p - C(x) > \frac{1}{\delta} \frac{d}{dx} \{(p - C(x))f(x)\} \quad \text{for} \quad \hat{x} < x < \bar{x}. \tag{34}$$

The method of proof will be to integrate this differential inequality along trajectories of the related equation

$$\frac{dx}{dt} = f(x); \tag{35}$$

these trajectories will be referred to as "natural trajectories." (This method has an intuitive economic basis that it would be out of place to describe in the present formal setting.)

For $0 \leq t \leq T$ we determine a map (see the Figure)

$$t \rightsquigarrow \tilde{t}$$

by following the natural trajectory beginning at the point $(t, x_{\hat{h}}(t))$ until it meets the graph of $x_h(t)$ at the point $(\tilde{t}, x_h(\tilde{t}))$. Note that $0 \to 0$ and $T \to T$. Equation

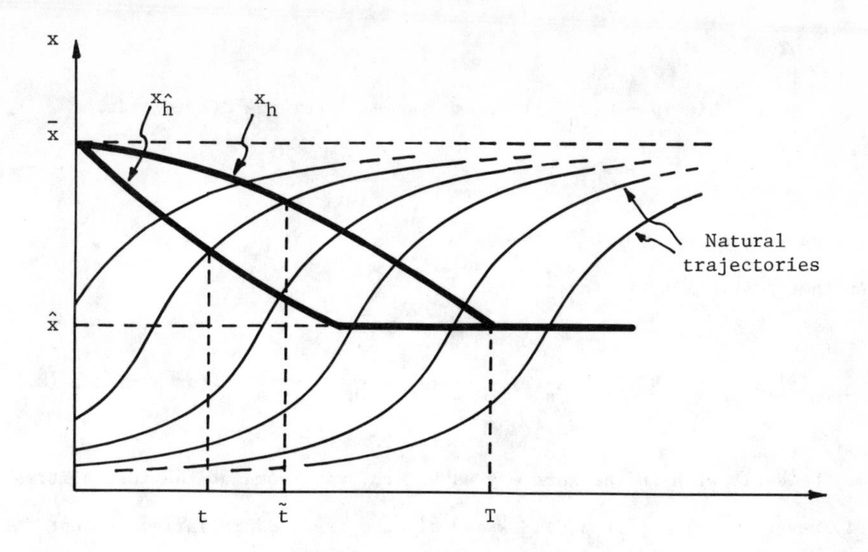

(35) implies that

$$\int_{x_{\hat{h}}(t)}^{x_h(\tilde{t})} \frac{dx}{f(x)} = \tilde{t} - t. \tag{36}$$

Integrating the inequality (34) from $x_{\hat{h}}(t)$ to $x_h(\tilde{t})$, we obtain

$$\{p - C(x_h(\tilde{t}))\}f(x_h(\tilde{t})) < \exp\left\{+\delta \int_{x_{\hat{h}}(t)}^{x_h(\tilde{t})} \frac{dx}{f(x)}\right\} \{p - C(x_{\hat{h}}(t))\}f(x_{\hat{h}}(t))$$

$$= e^{+\delta(\tilde{t}-t)}\{p - C(x_{\hat{h}}(t))\}f(x_{\hat{h}}(t)).$$

Also, it follows immediately from (36) that

$$\frac{h(\tilde{t})}{f(x_h(\tilde{t}))}\, d\tilde{t} = \frac{\hat{h}(t)}{f(x_{\hat{h}}(t))}\, dt.$$

Combining these formulas we finally obtain

$$\int_0^T e^{-\delta\tilde{t}} h(\tilde{t})\{p - C(x_h(\tilde{t}))\}d\tilde{t} < \int_0^T e^{-\delta t}\hat{h}(t)\{p - C(x_{\hat{h}}(t))\}dt.$$

This proves inequality (33).

Another possibility is the case

$$x_h(t) \leq x_{\hat{h}}(t) \quad (0 \leq t \leq T), \quad x_h(T) = x_{\hat{h}}(T), \quad x_h(t) < x_{\hat{h}}(t) \text{ for some } t \in (0,T).$$

This may be dealt with in the same way as before, with some of the inequalities suitably reversed. The case $x(0) \leq \hat{x}$ may also be treated similarly. Except for the details, the proof of the theorem is complete.

An interesting case (practically speaking) arises when equation (33) has no solution. In this case we may define $\hat{x} = 0$; then (34) holds and the given proof remains valid. Much less can be deduced if equation (33) has multiple solutions.

REFERENCES

1. Beverton, R.J.H., and Holt, S.J., On the Dynamics of Exploited Fish Populations. Ministry of Agriculture, Fisheries and Food (London), Fish. Inv. Ser. 2, Vol. 19 (1957).
2. Ciriacy-Wantrup, S.V., Resource Conservation: Economics and Policies. Univ. of California Press (Berkeley) (1963).
3. Clark, C.W., Profit maximization and the extinction of animal species, Journal of Political Economy 81, 950-961 (1973a).
4. _____, The economics of overexploitation, Science 181, 630-634 (1973b).
5. _____, When should whaling resume? (To appear) (1973c).
6. _____, Edwards, G., and Friedlaender, M., Beverton-Holt model of a commercial fishery: optimal dynamics, J. Fish. Research Board of Canada 30, 1629-1640 (1973).

7. Gordon, H.S., Economic theory of a common property resource: the fishery, Journal of Political Economy 62, 124-142 (1954).

8. Pontrjagin, L.S., et al, The Mathematical Theory of Optimal Processes, Pergamon Press (Oxford) (1964).

MODELS OF LARGE-SCALE NERVOUS ACTIVITY

J. D. Cowan

Department of Biophysics and Theoretical Biology

University of Chicago

Abstract

Consequences of a theory based on simple properties of excitatory and inhibitory nerve cells spatially organized into sheets are examined. Because of spatial organization, such sheets respond maximally to a limited range of optical contrast. Because of the temporal properties of individual cells, such sheets respond only to a limited range of temporal variations in stimuli. Because of the threshold properties of nerve cells, the responses themselves exhibit stable sub- and super-threshold behaviours. These properties are sufficient to build a theory for the simpler aspects of sensory perception and for the classification of regions of the brain in terms of their dynamical properties. Finally, for more complicated neural phenomena involving changes in neural responses and sensory adaptation, an extended theory involving synaptic plasticity is described and applied to various sensory phenomena.

References

H. R. Wilson and J. D. Cowan. Biophysical Journal 12 (1), 1972.

H. R. Wilson and J. D. Cowan. Kybernetik 13 (2), 1973.

See also M. I. T. Neurosciences Research Program Bulletin 12 (1), 1974.

Summary of

"STRATEGY FOR A PREDATOR WITH A MIMICKING PREY"

Presented at
Conference in Some Mathematical Problems in Biology 1973
Manuscript submitted to
The American Naturalist, 18 June 1973
by
G. F. Estabrook*
D. C. Jespersen**

A mathematical formulation of the phenomenon of close mimicry (Fisher 1930), incorporating the assumptions of Brower et al (1970), Huheey (1964), and Arnold (unpublished), including the concepts of relative abundance of models and mimics, relative noxiousness of the model, Markov process description of the spatial distributions of models and mimics, single trial predator learning, and deterministic ignoring behaviour can be structured as follows:

M = model (prey item that is bad to eat)

X = mimic (prey item that is good to eat).

Before actually eating a model or a mimic, the two cannot be distinguished by the predator. The hunting activity of the predator consists of a series, E_i, of encounters with models or mimics, so that $E_i = M$ or $E_i = X$ for $i = 1,2, 3,...$ Let us suppose, after S. Arnold, that the models and mimics are distributed in the predator's environment in such a way that the series E_i is a first order stationary Markov process. Denote the defining parameters of this process as follows:

$$p = Pr[E_i = M/E_{i-1} = X] ,$$

$$q = Pr[E_i = X/E_{i-1} = M] .$$

The transition matrix for this process is thus:

$$T = \begin{bmatrix} 1-p & p \\ q & 1-q \end{bmatrix}.$$

Let b quantify the relative badness of the model as follows:

b = (loss to predator from eating M)/(benefit to predator from eating X) .

*Department of Botany, U. of M., Ann Arbor, Mi. 48104
**Department of Mathematics, U. of M., Ann Arbor, Mi. 48104

After Huheey (1964) and Brower et al (1970), let us suppose that the predator continues to eat all the prey he encounters until he eats a model after which he ignores the next N prey he encounters (both models and mimics alike) before he begins to eat again.

We now have sufficient constraints to uniquely determine ν_1, the long term probability of eating a mimic on any given encounter, and ν_2, the long term probability of eating a model on any given encounter, as functions of p, q and N. For fixed p, q, and b, we may consider N to be the predator's strategy. The expected average long term benefit per encounter to the predator is thus

$$S(p\ ,\ q\ ,\ b\ ,\ N) = \nu_1 - b\nu_2 .$$

As shown by Estabrook and Jespersen (the manuscript here in summary)

$$\nu_1 = \frac{\frac{q}{p+q}(1-\alpha^{N+1})}{p(N+1) + \frac{q}{p+q}(1-\alpha^{N+1})},$$

and

$$\nu_2 = \frac{p}{p(N+1) + \frac{q}{p+q}(1-\alpha^{N+1})},$$

where $\alpha = 1 - p - q$.

We may now write

$$S(p\ ,\ q\ ,\ b\ ,\ N) = \frac{\frac{q}{p+q}(1-\alpha^{N+1}) - bp}{p(N+1) + \frac{q}{p+q}(1-\alpha^{N+1})}.$$

It is now straightforward to explore parameter space to discover those regions that support mimicry in the presence of a well-adapted predator. Some general conclusions suggest themselves immediately.

Given a large enough population of prey, an arbitrarily large proportion of mimics can be supported if the model is sufficiently noxious. In particular if

$$b > \frac{2q}{p(p+q)}$$

then $S < 0$ for all N and it is optimal for the predator to eat nothing.

For $b > \frac{q}{p(p+q)}$, $S < 0$ for "large" N . If $N = 0$, $S = \frac{q - bp}{p + q}$ so that when $b < \frac{p}{q}$ "eat nothing" is never optimal. For the special case of M and X independently distributed in space, i.e. $p + q = 1$, $S = \frac{q - bp}{p(N + 1) + q}$. In this case $b < \frac{q}{p}$ implies "eat everything" optimal, while $b > \frac{q}{p}$ implies "eat nothing" optimal.

The relatively small amounts of parameter space that correspond to large finite optimal strategies, suggest that the relatively large values for N reported in the literature, e.g. Brower (1969), evidence either "noisey" $N = \infty$ strategies, or the need for a more sophisticated mechanism for predator learning.

Literature Cited

Arnold, S. The evolution of modifiable behavior in relation to environmental pattern. Ms in preparation.

Brower, L. P. (1969) Ecological chemistry, Sci. Amer. 220 : 22-29.

Brower, L. P., Pough, F. H., and Meck, H. R. (1970) Theoretical investigations of automimicry, I. Single trial learning. Proc. Nat. Acad. Sci. 66 : 1059-1066.

Fisher, R. A. (1930) The Genetical Theory of Natural Selection. Oxford University Press, Oxford.

Huheey, J. E. (1964) Studies of warning coloration and mimicry IV. A mathematical model of model-mimic frequencies. Ecology 45 : 185-188.

FINITE DIFFERENCE METHODS FOR THE NUMERICAL SOLUTION OF THE NERNST-PLANCK-POISSON EQUATIONS

Robert J. French
Department of Zoology
Washington State University
Pullman, Washington 99163

INTRODUCTION

In this paper I shall describe two methods for the numerical solution of the kinetic equations for electrolyte diffusion and suggest conditions under which these methods might be applied to advantage in studies of biological systems.

Walther Nernst, in 1889, suggested that the movement of ionic particles in solution was determined by the sum of the forces due to gradients of concentration and electrical potential in the solution. In the next year, Max Planck observed that, as well as a flux equation for each ion present, one further relationship was necessary to solve for the ionic concentrations and electrical potential in a diffusion system. He pointed out that the electric charge density could be expressed as the sum of the charges on each ion present and thus incorporated Poisson's equation of electrostatics, which relates the electric potential to the charge density, into the system of equations to be solved. The term "Nernst-Planck equations" is usually applied to the flux equations only, even though Planck's original contribution was to add Poisson's equation to the system and suggest a means of solution.

Equations (1) show the Nernst-Planck-Poisson system of equations for a univalent binary electrolyte.

$$J = -\frac{dA}{dx} + A\frac{dV}{dx}$$

$$J\frac{D_a}{D_c} = -\frac{dC}{dx} - C\frac{dV}{dx}$$

$$\frac{d^2V}{dx^2} = \alpha(A - C) \qquad (1)$$

where

$$\alpha = L^2\frac{\tilde{A}(0)F^2}{\varepsilon RT} = \frac{L^2}{2L_D^2} \qquad (\varepsilon = \text{permittivity},\ L_D = \text{Debye length})$$

and the dimensionless variables are

$$A(x) = \tilde{A}(\tilde{x})/\tilde{A}(0)$$

$$C(x) = \tilde{C}(\tilde{x})/\tilde{A}(0)$$

$$J(x) = \tilde{J}(\tilde{x})\frac{L}{\tilde{A}(0)D_a}$$

$$V(x) = \tilde{V}(\tilde{x}) \frac{F}{RT}$$

$$x = \tilde{x}/L \, .$$

Here the variables are expressed in dimensionless units, A, C, V, J and x being the dimensionless anion and cation concentrations, voltage, anion flux and distance respectively. D_a and D_c are the anion and cation diffusion coefficients. Gradients in concentration and voltage are considered to exist in the x direction only. Nondimensionalization was carried out by forming the ratio between each variable expressed in normal laboratory units and some standard quantity in the same units. For the concentrations, for example, $\tilde{A}(0)$, the anion concentration at the beginning of the diffusion zone, was used. The choice of a standardizing quantity is arbitrary, however. RT/F, where R is the gas constant, T is the absolute temperature and F is the Faraday, is approximately 25.7 mv at T = 298°K.

The system of equations is first order in each concentration and is second-order in voltage. And it is decidedly non-linear. Under some circumstances shooting methods may be used to obtain solutions, but use of these methods frequently leads to instabilities. Wall and Berkowitz (1957) and Offner (1970) have reported such difficulties while attempting to solve similar problems. Consequently, I am directing my attention now to finite difference equation methods. Shooting methods require that some initial values not provided by the boundary conditions must be guessed, the equations then integrated and the final values checked against the known boundary conditions. Finite difference methods, in contrast, allow incorporation of all the boundary conditions into the difference representation of the differential equations. This presumably contributes to the greater stability of these methods during the iterative approach to a solution.

Figure 1 illustrates the physical problem considered in the initial investigation of the two methods that I shall describe. There is a diffusion zone of defined length between two solutions containing the same binary, univalent electrolyte at different concentrations. It was assumed that the boundary solution concentrations did not change and that the diffusion fluxes had reached a steady state--that is, that anion and cation fluxes were equal, and were constant in distance across the junction.

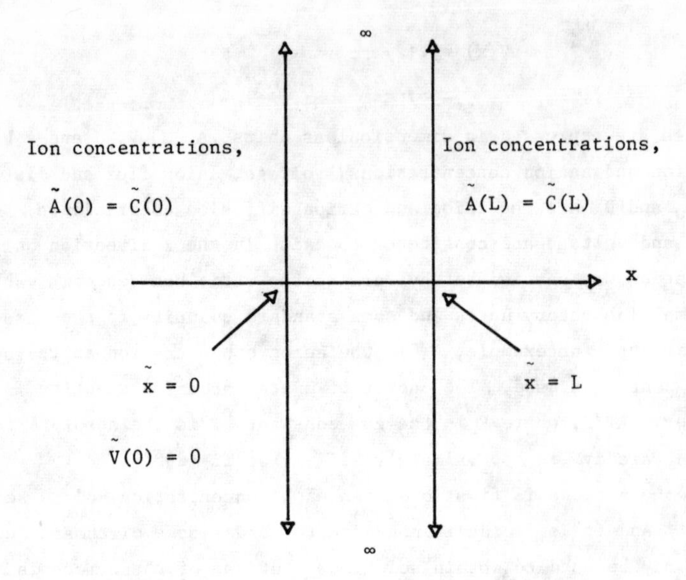

Figure 1. Physical model for the calculations.

METHODS

Generalized Newton's Method

In order to apply a generalized Newton's method, one equation was added to the system that was shown previously to give the vector equation (2).

$$Z' = F[Z(x)] = \begin{bmatrix} A' \\ C' \\ V'' \\ J' \end{bmatrix} = \begin{bmatrix} z_1' \\ z_2' \\ z_3' \\ z_4' \end{bmatrix} = \begin{bmatrix} -z_4 + z_1 z_3 \\ -z_4 D_a/D_c - z_2 z_3 \\ \alpha(z_1 - z_2) \\ 0 \end{bmatrix} \quad (2).$$

The unknown constant flux, J or z_4 is considered to be a variable and thus appears in the solution. After solving this system, the voltage was obtained by integrating the field across the interval using a standard Simpson's rule subroutine.

To start the procedure, one makes an initial guess at the concentration and voltage profiles and flux across the junction, and then applies the iterative relation shown below

$$Z'^{(n+1)} = Z'^{(n)} + \left(\frac{\partial F}{\partial Z}\right)^{(n)} \left(Z^{(n+1)} - Z^{(n)}\right) \quad (3)$$

where $\frac{\partial F}{\partial Z}$ denotes the Jacobian matrix of F with respect to changing Z. The ij^{th} element of the Jacobian matrix is given by

$$\left(\frac{\partial F}{\partial Z}\right)_{ij} = \frac{\partial}{\partial z_j}(Z_i')$$

for all i=1,...4 and all j=1,...4. Here the superscripts denote the level of iteration. The derivative vector $Z'^{(n+1)}$ may be approximated by a difference expression and an equation is thus obtained relating the new approximation to the solution, $Z^{(n+1)}$, to the previous approximation $Z^{(n)}$.

The form of the resulting matrix equation is

$$\begin{bmatrix} I & V_1 & 0 & \cdots & & 0 \\ 0 & I & V_2 & \cdots 0 \cdots & & 0 \\ \cdots & & & & & \cdot \\ 0 & \cdots & 0 & I & V_{M-1} & 0 \\ 0 & \cdots & & 0 & I & V_M \\ B_0 & 0 & \cdots & & 0 & B_L \end{bmatrix} \cdot \begin{bmatrix} Z_1^{(n+1)} \\ Z_2^{(n+1)} \\ \vdots \\ Z_{M-1}^{(n+1)} \\ Z_M^{(n+1)} \\ Z_{M+1}^{(n+1)} \end{bmatrix} = \begin{bmatrix} U_1 \\ U_2 \\ \vdots \\ U_{M-1} \\ U_M \\ G \end{bmatrix} \quad (4)$$

where the boundary conditions are expressed by the matrix equation

$$B_0 Z_1^{(n+1)} + B_L Z_{M+1}^{(n+1)} = G .$$

The blocks $Z_i^{(n+1)}$ are four component column-vectors representing the two concentrations, the voltage and the flux for the grid points i at the (n+1)th iteration. Four boundary conditions are specified by the vector block G and are inserted as initial or final values for the appropriate variable by choosing suitable elements for the matrix blocks B_0 and B_L. In principle, any combination of four boundary conditions may be used equally easily, and this flexibility was one of the attractive features of the method. Another was that the procedure did converge to give solutions for a wide range of problems. Some of these solutions are to be discussed later in this talk.

As most of the off-diagonal elements in the coefficient matrix are zero an efficient programme for the solution of the system of equations may be written. The largest general matrix (i.e. all elements allowed to be non-zero) that must be inverted is a 4 x 4.

A Point Iterative Method

The second method which I shall describe is considerably less tedious to program, and although it did not converge to provide solutions over as wide a range as the Newton's method, it did provide some solutions under conditions for which the Newton's method routine did not converge.

This procedure was developed with a view to extending the calculations to flows in more than one space dimension, and a different form of the equations was used. The differential form, (5), and the difference form, (6), of the equations appear below.

$$0 = -\frac{d^2A}{dx^2} + A\frac{d^2V}{dx^2} + \frac{dA}{dx}\frac{dV}{dx}$$

$$0 = -\frac{d^2C}{dx^2} - C\frac{d^2V}{dx^2} - \frac{dC}{dx}\frac{dV}{dx}$$

$$\frac{d^2V}{dx^2} = \alpha\ (A - C) \qquad (5).$$

$$0 = -(A_{m+1} - 2A_m + A_{m-1})/h^2 + A_m(V_{m+1} - 2V_m + V_{m-1})/h^2$$
$$+ (A_{m+1} - A_{m-1})(V_{m+1} - V_{m-1})/4h^2$$

$$0 = -(C_{m+1} - 2C_m + C_{m-1})/h^2 - C_m(V_{m+1} - 2V_m + V_{m-1})/h^2$$
$$- (C_{m+1} - C_{m-1})(V_{m+1} - V_{m-1})/4h^2$$

$$\alpha(A_m - C_m) = (V_{m+1} - 2V_m + V_{m-1})/h^2 \qquad (6).$$

Here one solves for the concentrations and the electric potential directly. The equations for the concentration were obtained by differentiation of the Nernst-Planck flux equations under the steady-state assumption which implies, in one dimension, that $dJ/dx = 0$. Note that solution of the analogous equations in 2 or 3 dimensions would require solution only for the scalars concentration and voltage. The components of the flux and field vectors, if required, could then be obtained from the gradients of the concentrations and potential.

For convenient application of a point-iterative procedure one wishes to be able to solve the difference equation for one variable at a given point explicitly in terms of its value at neighbouring points and of the values of the other variables. Notice that in the difference form of the equations, the upper two may easily be rearranged to solve for the concentrations at the central gridpoint provided that an estimate of the other values is available. The voltage may be calculated from the lowest difference equation in the same way. Thus, given a reasonable initial guess at concentration and voltage profiles, one can cycle through the grid points calculating new values for each variable until the change from one iteration to the next becomes insignificant.

A few practical details proved important in the application of this method. Convergence was only obtained when the new values for voltage at each iteration level were obtained in a separate calculation loop from that in which new concentration values were obtained. Within each of the two calculation loops successive displacement was used without ill effect on convergence. That is, new concentrations for each point were inserted into the arrays as soon as calculated, and were then used for calculations at all subsequent points. The need to calculate the voltage in a separate loop may be due to the fact that the voltage is determined by small differences in concentrations at each point so that the procedure is only sufficiently stable to converge if the complete set of concentrations at one iteration is consistent with a complete set of potential values over the whole grid of points.

Faster convergence was obtained by use of a successive over-relaxation procedure. Rather than taking as the new iterate at a point the value obtained simply by solving the difference equation as indicated above, one projects beyond that value by an amount fixed by the choice of a relaxation parameter r. Rearranging the difference equations one obtains

$$\bar{A}_m = [(A_{m+1} + A_{m-1})/h^2 - (A_{m+1} - A_{m-1})(V_{m+1} - V_{m-1})/4h^2]$$

$$/[2/h^2 + (V_{m+1} - 2V_m + V_{m-1})/h^2]$$

$$\bar{C}_m = [(C_{m+1} + C_{m-1})/h^2 + (C_{m+1} - C_{m-1})(V_{m+1} - V_{m-1})/4h^2]$$

$$/[2/h^2 - (V_{m+1} - 2V_m + V_{m-1})/h^2]$$

$$\bar{V}_m = - [\alpha(A_m - C_m)h^2 - (V_{m+1} + V_{m-1})]/2 \qquad (7).$$

The new values to be inserted at point m in the arrays are given by the following expressions

$$V_m^{(n+1)} = V_m^{(n)} + r\,(\bar{V}_m - V_m^{(n)})$$

$$A_m^{(n+1)} = A_m^{(n)} + r\,(\bar{A}_m - A_m^{(n)})$$

$$C_m^{(n+1)} = C_m^{(n)} + r\,(\bar{C}_m - C_m^{(n)}) \qquad (8).$$

The quantities $\bar{A}_m$, $\bar{C}_m$ and $\bar{V}_m$ are obtained simply by rearranging the difference equations. The new approximation is then obtained from the relaxation expressions below. Values of 1.2-1.5 for r produced the best results for one dimensional problems.

In the initial description of the physical problem, 5 boundary conditions were given--the concentrations at each end of the junction and the voltage (defined as zero) at one end. For this method a system of three second-order equations was used, and thus one further boundary condition was required. The three equations imply no particular relationship between the fluxes of anion and cation. Indeed fluxes no longer appear explicitly. Hence, the flux, J, may be eliminated from the Nernst-Planck equations (the flux relations in the equations (1)) to yield an expression for the voltage gradient in terms of the concentrations.

$$\frac{dV}{dx} = \frac{\left(\frac{dA}{dx} - \frac{dC}{dx}\frac{D_c}{D_a}\right)}{\left(A + C\,\frac{D_c}{D_a}\right)} \qquad (9).$$

The difference form of this equation provides the required boundary condition.

This relationship holds for all points in the junction. In particular, it may be used to provide a floating boundary condition on the voltage for each iteration, calculated from the current concentration values. For convergence it was necessary to calculate the second boundary condition at the opposite end of the junction from that at which the voltage was defined as zero.

RESULTS AND DISCUSSION

Analytic Approximations

Two simplifying assumptions have been applied to the Nernst-Planck-Poisson system in order to obtain solutions. The first, introduced by Planck, is that for practical purposes electroneutrality holds at each point in the diffusion zone, or for a 1:1 electrolyte, anion and cation concentrations are equal to each other. The other assumption, used by Goldman in 1943, is that the electric field in the junction is constant. Both approaches, in essence, replace Poisson's equation by a simpler relation, then go about the integration of the flux equations.

The results obtained using these two assumptions to predict the diffusion potential between two solutions of a univalent 1:1 electrolyte are given below. From the electroneutrality assumption,

$$\tilde{V}_2 - \tilde{V}_1 = \frac{RT}{F}\frac{w-u}{w+u}\ln\frac{C_2}{C_1} \qquad (10),$$

and from the constant field assumption,

$$\tilde{V}_2 - \tilde{V}_1 = \frac{RT}{F}\ln\frac{uC_1 + wC_2}{uC_2 + wC_1} \qquad (11).$$

u, w are cation and anion mobilities. I would like to point out some common features. First, both predictions are independent of junction thickness. Second, both predictions are dependent only on relative concentrations and relative mobilities of the ions, not absolute values of either.

In Figure 2 concentration and voltage profiles obtained by numerical integration of the Nernst-Planck-Poisson equations for two junction thicknesses are shown.

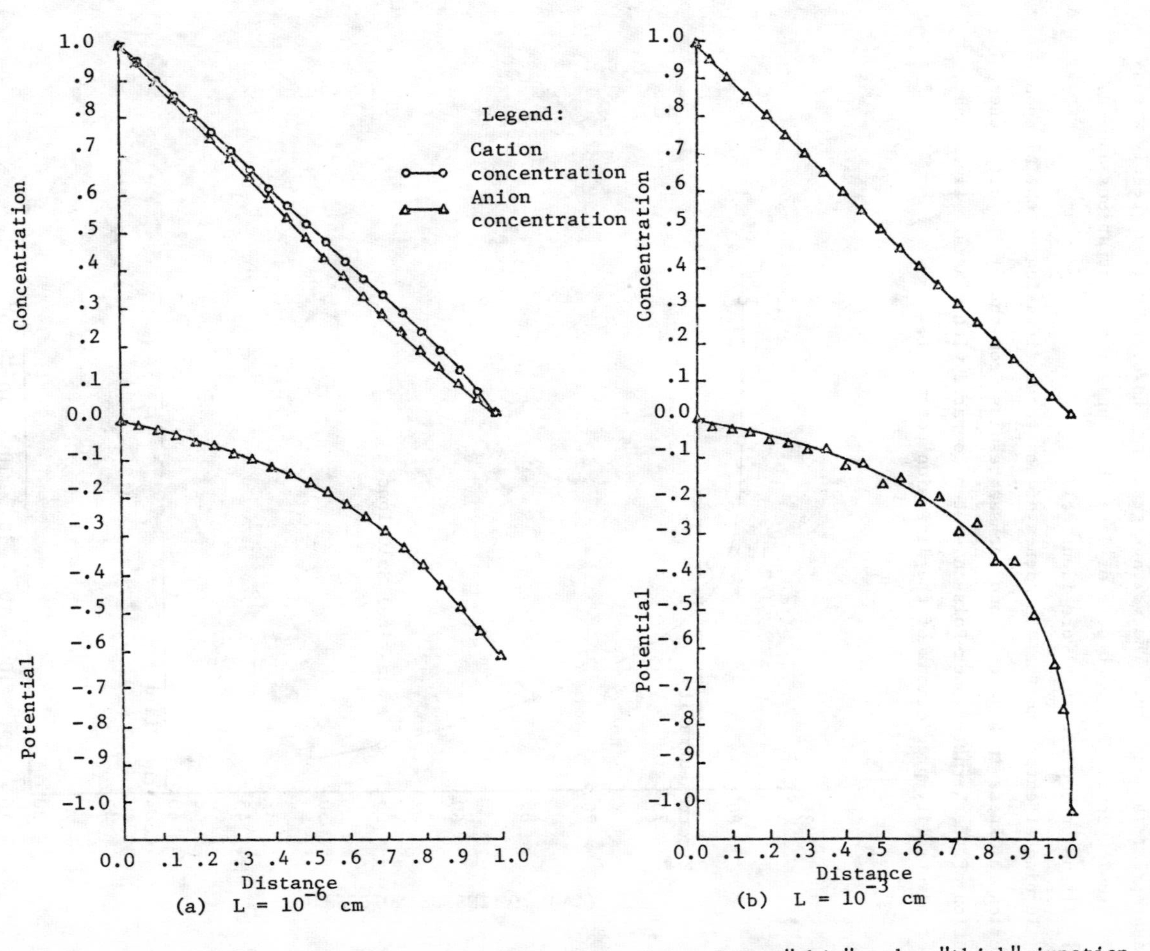

Figure 2. Concentration and electric potential profiles for a "thin" and a "thick" junction.

For the thinner junction (10^{-6} cm), the anion and cation concentrations are shown clearly separated, implying that electroneutrality would be a poor approximation. There is also perceptible curvature of the potential profile--the electric field is not constant. For the thicker junction (10^{-3} cm), the anion and cation concentrations are not distinguishable--here, electroneutrality would have been an appropriate assumption. The voltage gradient, however, varies greatly across the diffusion zone. For both sets of data the boundary concentrations and the ionic diffusion coefficients were identical.

Turning attention to the dependence of the total diffusion potential upon junction thickness in Figure 3, one can see results from the finite difference solution of the Nernst-Planck-Poisson system compared with constant field and electroneutrality predictions of the diffusion potential.

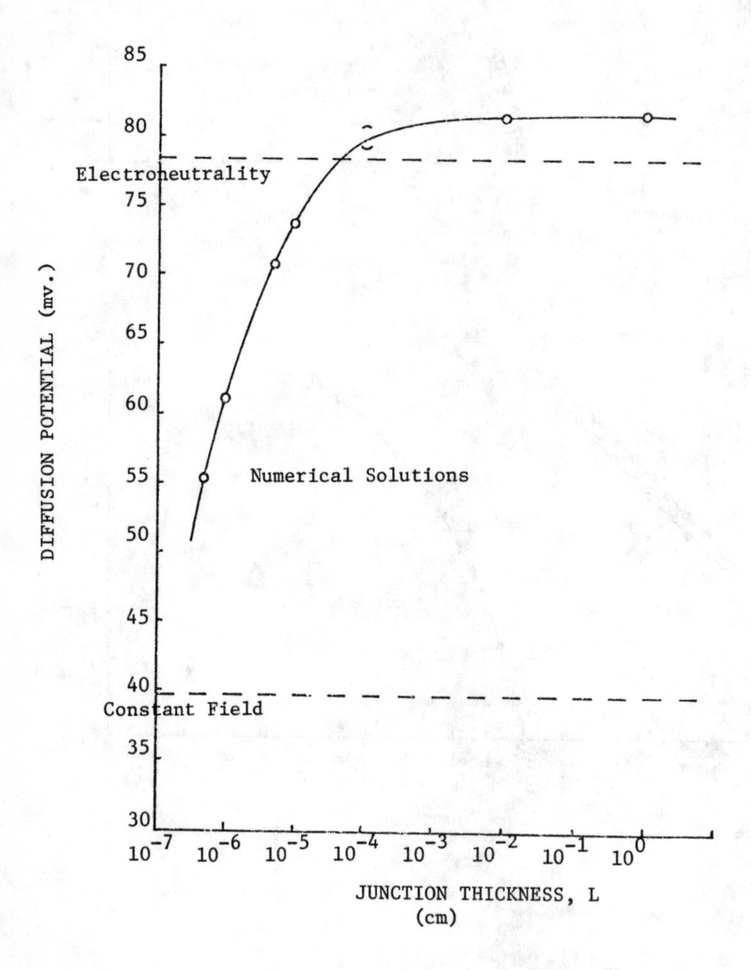

Figure 3. Dependence of predicted diffusion potential on junction thickness.

The numerical solutions show a marked dependence on the junction thickness, varying from values approaching the constant field prediction for thin junctions to values somewhat higher than the electroneutrality prediction for the thickest junctions. The range considered varies from about plasma membrane thickness to the macroscopic dimensions of some ion exchange membranes that are used in experimental studies. The concentration difference across the junction is constant for these calculations, A(0) being 0.5 M-lit^{-1} and A(L) being 0.005 M-lit^{-1}. The ratio of the diffusion coefficients was about 4.9, the cation being more mobile.

Figure 4 depicts calculations for which a ten-fold concentration ratio was assumed, but the absolute concentrations varied.

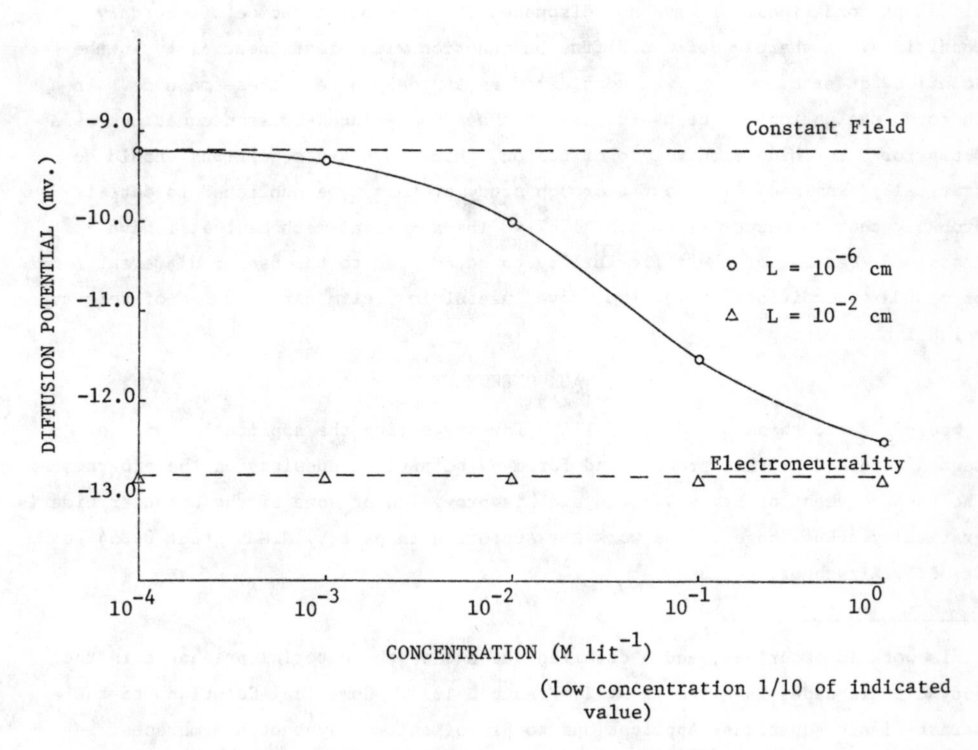

Figure 4. Concentration dependence of predicted diffusion potential (constant concentration ratio)

The concentration range spans values that are commonly employed in experimental studies on synthetic and biological membranes. The ratio of cation mobility to anion mobility was 0.78. Predictions for the thick membrane from solution of the complete Nernst-Planck-Poisson system closely follow the values from the electroneutrality assumption. For the thin membrane, however, numerical solutions are approximated at one extreme (low concentrations) by the constant field prediction and at the other (high concentrations) by the electroneutrality value. Between these extremes lies a range of concentrations for which neither of the approximations is particularly good.

Although it has been possible to predict qualitatively conditions under which the constant field and electroneutrality assumptions are likely to be poor approximations (Bass, 1964; MacGillivray, 1968; MacGillivray and Hare, 1969; Arndt, Bond and Roper, 1972) actual deviations from the solutions of the complete Nernst-Planck-Poisson system can only be evaluated by a method capable of solving the complete system of equations.

Finally, I should point out that in this paper I have considered only methods for integration of the Nernst-Planck-Poisson system, and their appropriateness under different conditions. I have not discussed the particular choice of boundary conditions. A sharply defined diffusion junction with electroneutrality in the solutions at each boundary was considered as a model to test these methods. In a thorough evaluation of the usefulness of the Nernst-Planck-Poisson equations as a means for describing electrolyte diffusion, these boundary conditions should be critically examined. The most thorough study of this type published to date is probably that by Bruner (1965,a,b; 1967). The numerical methods that I have described, because of their flexibility to be adapted to the use of different forms of boundary conditions, have also proved useful in preliminary studies of this type which I have made[1].

ACKNOWLEGMENT

I would like to thank Dr. J.L. Phillips for suggesting the application of the Newton's method to this problem and for many helpful discussions on the programming. The encouragement of Dr. G.W. Swan and his provision of some of the computer time is gratefully acknowledged. The work was supported in part by NIGMS grant 04254 to Dr. L.B. Kirschner.

---

[1]This work is described, and a detailed account of the material presented in the body of this paper is given in R. J. French. 1973. Numerical Solutions to the Nernst-Planck Equations--Applications to Biological and Synthetic Membranes. Ph.D. Thesis, Washington State University.

BIBLIOGRAPHY

Arndt, R. A., J. D. Bond, and L. D. Roper. 1972. Electroneutral approximate solutions of the steady state electrodiffusion equations for a simple membrane. J. Theoret. Biol. 34: 265-276.

Bass, L. 1964. Potential of liquid junctions. Trans. Faraday Soc. 60: 1914-1918.

Bruner, L. J. 1965a. The electrical conductance of semipermeable membranes. I. A formal analysis. Biophys. J. 5: 867-886.

Bruner, L. J. 1965b. The electrical conductance of semipermeable membranes. II. Unipolar flow, symmetric electrolytes. Biophys. J. 5: 887-908.

Bruner, L. J. 1967. The electrical conductance of semipermeable membranes. III. Bipolar flow--symmetric electrolytes. Biophys. J. 7: 947-972.

Goldman, D. E. 1943. Potential, impedance and rectification in membranes. J. Gen. Physiol. 27: 37-60.

MacGillivray, A. D. 1968. Nernst-Planck equations and the electroneutrality and Donnan equilibrium assumptions. J. Chem. Phys. 48: 2903-2907.

MacGillivray, A. D., and D. Hare. 1969. Applicability of Goldman's constant field assumption in biological systems. J. Theoret. Biol. 25: 113-126.

Nernst, W. 1889. Die elektromotorische Wirksamkeit der Ionen. Z. Phys. Chem. 4: 129-181.

Offner, F. F. 1970. Kinetics of excitable membranes. Voltage amplification in a diffusion regime. J. Gen. Physiol. 56: 272-296.

Planck, M. 1890a. Ueber die Erregung von Elektricität und Wärme in Elektrolyten. Ann. Phys. Chem. N.F. 39: 161-186.

Planck, M. 1890b. Ueber die Potentialdifferenz zwischen zwei verdünnten Lösungen binarer Elektrolyte. Ann. Phys. Chem. N.F. 40: 561-576.

Wall, F. T., and J. Berkowitz. 1957. Numerical solution of the Poisson-Boltzmann equation for spherical polyelectrolyte molecules. J. Chem. Phys. 26: 114-122.

HABITAT SELECTION AND A LIAPUNOV FUNCTION
FOR COMPETITION COMMUNITIES

by

Michael E. Gilpin
Department of Biology
University of California
at San Diego
La Jolla, California 92307

Game theory, though it was invented with human economic behavior in mind, has provided valuable insights in the field of population biology. For example, Trivers (1971) has investigated the evolution of "altruistic" behavior by a game theoretic argument. And Maynard Smith (1972) has similarly analyzed the tactical possibilities of agonistic encounters.

Game theory has had success because the pay-off in all evolutionary or ecological games--whether between alleles, individuals, populations or species--is fitness. But this, as Slobodkin (1964) has pointed out, gives such games an existential character, for fitness cannot long average above unity, but it can go to zero. There is no winning, only losing. To win is simply to go on playing; but to lose is to go extinct, and this is irreversible.

Using a Liapunov function I have constructed (Gilpin, 1974), which gives information on the probability of population extinction between interspecific competitors, I shall in this paper analyze the strategies that should be followed in a "habitat selection" game in which taxonomically similar species restrict their exploitation of the environment in hopes of improving their long term survival probabilities.

The Liapunov function I have constructed is algorithmically defined and can be applied to any differential equations that describe interspecies competition. For a completely symmetrical, two species Lotka-Volterra competition community, which has the dynamic equations

$$\frac{dN_i}{dt} = \frac{rN_i}{K}(K - N_i - aN_j),\ i=1,2,\ i\neq j, \qquad (1)$$

where *r* is the intrinsic rate of increase of a population, *K* is the saturation density (carrying capacity) of a single population, and *a* is the competition coefficient (the "niche overlap") between populations, it is possible to derive an analytical approximation that gives the value of the Liapunov function, V^{ext}, at a point of one species extinction. In the presence of environmental disturbances, the extinction probability of one or the other of the two species is inversely related to this value: when V^{ext} is zero the probablity of extinction is unity, and when V^{ext} is very large the probability of extinction is small. The exact extinction probabilities depend on the nature and the magnitude of the disturbances.

The analytical approximation for the Liapunov function at one species points of extinction is

$$V^{ext} = \frac{rK^2 (1 - a^2)}{6(1 + a)^3}. \tag{2}$$

The parameter *r* in this expression will not be varied in the game that is to follow; it may be assumed to have the value unity. If *a* is greater than 1, V^{ext} is negative and the system is absolutely unstable. If *a* is less than 1 the system may still be more or less unstable in a disturbed environment.

Tha habitat selection game is assumed to be played symmetrically by the two species; that is, they each make the same move at the same time. Since the optimal strategy must be the same for the two species, this assumption is not unreasonable. The species move by selecting for their occupation a restricted subset (a habitat) of the available environment. This must necessarily reduce their single species carrying capacity, *K*. Such a reduction of *K* is also likely to have an effect on *a*, the amount of niche overlap between the species, since it is improbable that the two species would select exactly the same habitat.

The moves in this game may be evaluated with the use of equation (2). A move, a change in *K* and *a*, is good if and only if it increases the value of V^{ext} and thereby lowers the extinction probability.

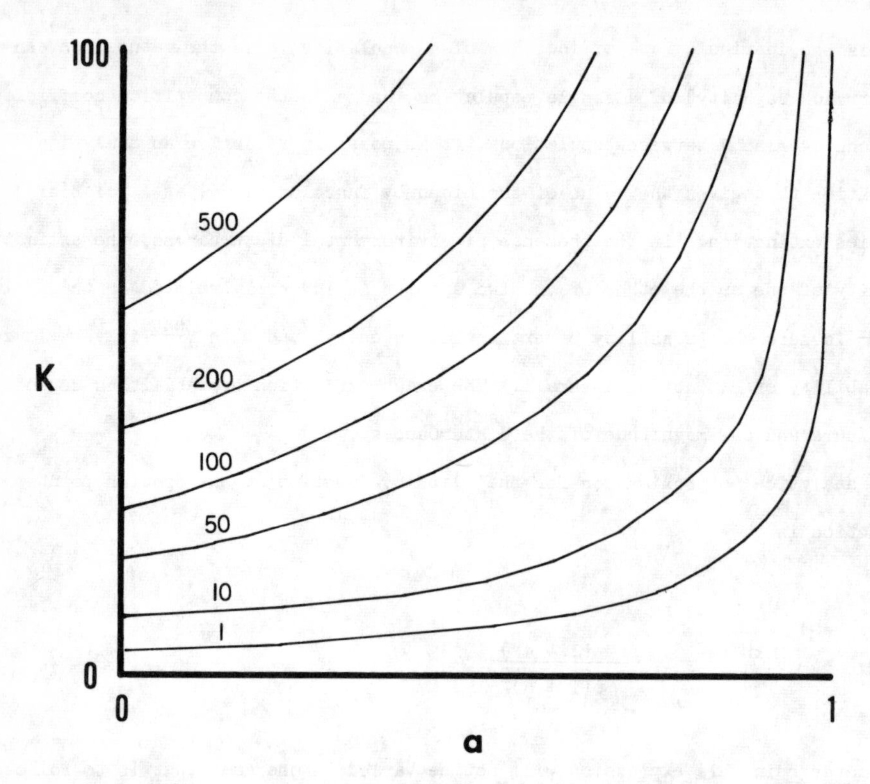

Figure 1. Lines of constant Liapunov extinction potential. The number above each line is the value $\underline{V}^{\underline{ext}}$ for that line.

Figure 1 shows lines of constant $\underline{V}^{\underline{ext}}$ plotted on a $\underline{K}$-$\underline{a}$ coordinate system. A move may be thought of as a vector on this coordinate space: if the vector ends at a higher $\underline{V}^{\underline{ext}}$ than it started at, it describes a good move.

Suppose the game began at a high $\underline{K}$ and a high $\underline{a}$, which would mean that the two species were competitors throughout most of the environment. For this case, the beginning of the move vector is at the upper right hand corner of Figure 1. It is obvious that a species could successfully select a very small habitat (reduce $\underline{K}$ greatly) so long as this was accompanied by a slight reduction of $\underline{a}$. Conversely, near the $\underline{K}$ axis (for small $\underline{a}$) only a very small reduction of $\underline{K}$ can be of benefit; no matter how large the reduction of $\underline{a}$.

This game can be generalized to include a restriction (or selection) of diet.

Such a restriction would reduce a species' carrying capacity and would likely at the same time reduce its niche overlap with its competitor. Thus, any reduction of $\underline{K}$ may be profitable if it simultaneously reduces $\underline{a}$.

To provide ancedotal support for this theory, consider the Hawaiian *Drosophila*. The fauna of these islands includes some 600 species of this genus. Since the islands are only a few million years old, speciation and adaptive radiation in this genus must have been rapid. Yet one wonders why competitive exclusion has not held the number of *Drosophila* species lower.

In *Drosophila*, as in many insect taxa, interspecies competition occurs almost completely between larvae; the adults are "only for" gene exchange and dispersal and do not make demands on resources. Bill Heed (personal communication) has found that the larval niches for *Drosophila* are very narrow on the Hawaiian Islands. A species normally restricts oviposition much further than seems necessary on purely physiological grounds; that is, it selects a subset of the available larval environment. And a particular type of larval habitat is often occupied by but a single species. It thus appears that in the course of their evolutionary history, the Hawaiian *Drosophila* have played the habitat selection game and that the strategy they have adopted corresponds to that theoretically predicted.

References

Gilpin, M. E. (1974) A Liapunov Function for Competition Communities. *Journal of Theoretical Biology*. in press.

Slobodkin, L. B. (1964) The Strategy of Evolution. *American Scientist*: 52, 342.

Smith, J. M. (1972) *On Evolution*. Aldine, New York.

Trivers, R. L. (1971) The Evolution of Reciprocal Altruism. *Quarterly Review of Biology*: 46, 35.

ASPECTS OF EVOLUTIONARY THEORY AND THE THEORY OF GAMES

by

Roger I.C. Hansell and Ezio Marchi,
Dept. of Zoology and Dept. of Mathematics,
University of Toronto.

INTRODUCTION

Lewontin (1961) proposed that game theory be modified to form a predictive evolutionary calculus. The need for such a development depends on the wide disparity between the patterns of macroevolution and the results of population genetics. Analytic population models for single genes were developed by Fisher (1930) and Wright (1931). Recently Karlin and Feldman (1970) have developed the mathematical theory for two genes. Clearly the step to developing population genetic models for entire genotypes is a large one, particularly when the problem of genes acting through integrated phenotypes is considered.

In most game theory models, organisms with particular phenotypes act as players, interacting with each other and with the environment. Slobodkin (1967), and Rapaport (1970), developed the concept of existential games in which the payoff to the players is reduced solely to the ability to continue to play the game.

Another kind of model (Marchi and Hansell 1973a) deals with the pattern of evolution as indicated by parsimonious or minimal change evolutionary trees. The implications of the relaxation of the parsimony criterion are considered, as for example when climate changes to less stringent conditions for some subset of organisms, or when a new environment is invaded so that the likelihood of convergent or parallel evolution is increased. If the different organisms are seen as interacting with different aspects of the environment, different kinds of parsimony criteria may be employed, influencing the decision as to the optimal phylogenetic tree. This approach is generalized (Marchi and Hansell 1974) by incorporating the taxonomists evaluation of the relatedness of the organisms: the theory of fuzzy sets is applied to the phenotype character space, the partitions of the organisms, and as a consequence to the evolutionary tree.

Macroevolution has also been simulated (Marchi and Hansell 1973b) in a game model which considers the competitive interaction between populations in the process of speciating, with other groups of speciating populations, and with other established populations. The strategies of the population are given by the similarities between their phenotypes. The three kinds of interactions are associated with different mechanisms of transmitting information between the populations; for instance, information in the form of gene flow between speciating populations is shown as constraining the set of possible similarities of any one of the populations given the conditions of the others.

Applications of game theory in evolution may be criticized on the grounds that the selective process in evolution is intrinsically different from the minimax approach to a solution in games. This argument appears to stem from the idea that the minimax solution is only applicable to situations in which intelligent or goal seeking behaviour is involved.

As a first step towards demonstrating the general applicability of certain approaches of game theory to evolutionary problems, we will take a simple biological example of the process of selection and compare the results of the iterative process of selection with the minimax approach to the equivalent evolutionary game expressed in normal form. The concept of 'payoff' is used as a technique for determining the optimal strategies, not in the sense of an external value system.

ARGUMENT

Consider a set of species indexed by $N=\{1,\ldots,n\}$. For each species $i\varepsilon N$ let $N_i=\{1,\ldots,m_i\}$ be the set of all phenotypes which may be expressed by individuals of species i. An individual can express only one phenotype; we index the phenotype by $j_i \varepsilon N_i$.

The biomass of individuals with phenotype j_i of species $i\varepsilon N$ in generation t is given by $b^t_{ij_i}$. Then the vector $b^t = (b^t_1,\ldots,b^t_N)$ where $b^t_i = (b^t_{i1_i}, b^t_{i2_i},\ldots,b^t_{im_i})$, gives the total distribution of the population of species by phenotypes at time t. Let the total resource available in one generation be R energy units. Let the amount of resource obtainable per unit biomass of phenotype j_i of species i be r_{ij_i} when the resource is available in unlimited supply.

The efficiency of converting the resource into new biomass of phenotype j_i of species i is given by $0 \le \gamma_{ij_i} \le 1$. The natural fecundity per unit biomass of phenotype j_i of species i is given by C_{ij_i}.

Now we consider the simplest kind of competition, that in which there is no inhibition or interaction between organisms in the process of obtaining the resource and the amount of resource obtained is proportional to the demand for it. We can describe the demand for the resource in generation t as the potential biomass of that phenotype and species in that generation.

We consider the special case that the fecundity of a phenotype is of the same order of magnitude as the carrying capacity given by $R\,\gamma_{ij_i}$. It is reasonable to consider that the actual biomass of phenotype j_i species i in a given generation is proportional to the amount of resource obtained. Then the biomass of phenotype j_i species i in the generation t + 1 is given by:

$$b^{t+1}_{ij_i} = R \frac{b^t_{ij_i}\; r_{ij_i}\; \gamma_{ij_i}\; C_{ij_i}}{\sum_{k=1}^{n} \sum_{\ell_k=1}^{m_k} b^t_{k\ell_k}\; r_{k\ell_k}\; \gamma_{k\ell_k}\; C_{k\ell_k}} .$$

By induction, the general recursive formula at the generation t is

$$b^{t}_{ij_i} = R \frac{b^o_{ij_i}\; (r_{ij_i}\; \gamma_{ij_i}\; C_{ij_i})^t}{\sum_{k=1}^{n} \sum_{\ell_k=1}^{m_k} b^o_{k\ell_k}\; (r_{k\ell_k}\; \gamma_{k\ell_k}\; C_{k\ell_k})^t} .$$

Then

$$b^{t}_{ij_i} = R \frac{b^o_{ij_i}}{\sum_{k=1}^{n} \sum_{\ell_k=1}^{m_k} b^o_{k\ell_k} \left(\frac{r_{k\ell_k}\; \gamma_{k\ell_k}\; C_{k\ell_k}}{r_{ij_i}\; \gamma_{ij_i}\; C_{ij_i}}\right)^t} .$$

69

Consider the set of individuals with a maximal value of combined feeding capacity, efficiency and fecundity:

$$H = \{(ij_i): r_{ij_i} \gamma_{ij_i} C_{ij_i} = \max_{k\ell_k} r_{k\ell_k} \gamma_{k\ell_k} C_{k\ell_k}\} \neq \phi,$$

where ϕ is the empty set.

The expression $\left(\frac{r_{k\ell_k} \gamma_{k\ell_k} C_{k\ell_k}}{r_{ij_i} \gamma_{ij_i} C_{ij_i}}\right)^t$ as $t \to \infty$

takes the value 0 if $(k\ell_k) \notin H$ and $(ij_i) \in H$,

∞ if $(k\ell_k) \in H$ and $(ij_i) \notin H$,

1 if $(k\ell_k) \in H$ and $(ij_i) \in H$.

Substituting in above, if $(ij_i) \varepsilon H$ and no other $(k\ell_k) \varepsilon H$ then $b^t_{ij_i} \to R$. That is phenotype j_i of species i is the sole survivor of the selective process. If $(ij_i) \notin H$ then $b^t_{ij_i} \to 0$, that is phenotype j_i of species i goes to extinction. If $(ij_i) \varepsilon H$ and some other $(k\ell_k) \varepsilon H$, then $b^t_{ij_i} \to \dfrac{R\, b^o_{ij_i}}{\sum_{(k\ell_k)\varepsilon H} b^o_{k\ell_k}}$.

That is, an equilibrium condition is achieved between phenotypes with the maximal fitness.

In order to show the intuitive equivalue of game theory, we demonstrate that the results of the above iterative selective process are biologically equivalent to a game theory solution.

Let the payoff matrix be

$$A_i\ (j_1, \ldots, j_p) = \frac{r_{j_i} \gamma_{j_i} C_{j_i}}{\sum_{\ell \varepsilon P} r_{j_\ell} \gamma_{j_\ell} C_{j_\ell}}$$

where P is the set of surviving species $P \subset N$, that is those species whose possible strategies contain at least one which is a member of the maximum set. Now, consider a joint point $(\bar{j}_1, \ldots, \bar{j}_p)$ where $\bar{j}_i \varepsilon H_i = \{j_i: (ij_i) \varepsilon H\}$,

since $\dfrac{r_{j_i} \gamma_{j_i} C_{j_i}}{r_{\bar{j}_i} \gamma_{\bar{j}_i} C_{\bar{j}_i}} \leq 1$

then $$\frac{\sum_{\ell\varepsilon P-\{i\}} r_{\bar{j}_\ell}\gamma_{\bar{j}_\ell}C_{\bar{j}_\ell} + r_{j_i}\gamma_{j_i}C_{j_i}}{\sum_{\ell\varepsilon P-\{i\}} r_{\bar{j}_\ell}\gamma_{\bar{j}_\ell}C_{\bar{j}_\ell} + r_{\bar{j}_i}\gamma_{\bar{j}_i}C_{\bar{j}_i}} \le 1$$

and $$\frac{\sum_{\ell\varepsilon P-\{i\}} r_{\bar{j}_\ell}\gamma_{\bar{j}_\ell}C_{\bar{j}_\ell} + r_{j_i}\gamma_{j_i}C_{j_i}}{\sum_{\ell\varepsilon P-\{i\}} r_{\bar{j}_\ell}\gamma_{\bar{j}_\ell}C_{\bar{j}_\ell} + r_{\bar{j}_i}\gamma_{\bar{j}_i}C_{\bar{j}_i}} \ge \frac{r_{j_i}\gamma_{j_i}C_{j_i}}{r_{\bar{j}_i}\gamma_{\bar{j}_i}C_{\bar{j}_i}}$$

or $$\frac{r_{\bar{j}_i}\gamma_{\bar{j}_i}C_{\bar{j}_i}}{\sum_{\ell\varepsilon P} r_{\bar{j}_\ell}\gamma_{\bar{j}_\ell}C_{\bar{j}_\ell}} \ge \frac{r_{j_i}\gamma_{j_i}C_{j_i}}{\sum_{\ell\varepsilon P-\{i\}} r_{\bar{j}_i}\gamma_{\bar{j}_i}C_{\bar{j}_i} + r_{j_i}\gamma_{j_i}C_{j_i}} .$$

That is $A_i(\bar{j}_1,\ldots, \bar{j}_{i-1}, \bar{j}_i, \bar{j}_{i+1},\ldots \bar{j}_p) > A_i(\bar{j}_1,\ldots, \bar{j}_{i-1}, j_i, \bar{j}_{i+1},\ldots, \bar{j}_p)$

which gives the existence of the required equilibrium point.

We point out that in the case of two species, since the game is constant sum, the equilibrium point $(\bar{j}_1, \bar{j}_2)$ is a saddle point for the matrix A_1, fulfilling $A_1(j_1, j_2) = \max_{j_1} \min_{j_2} A_1(j_1,j_2) = \min_{j_2} \max_{j_1} A_1(j_1,j_2)$, hence j_1 is a maximin and j_2 is a minimax solution respectively.

We further point out that where these equilibrium points are biologically equivalent, technically the expression can be normalized to an equilibrium point in the mixed strategy, having the proportional payoffs as any equilibrium point in pure strategy, since

$$\sum_i \sum_{j_i} \frac{b^t_{ij_i}}{R} = 1$$

for each t>1.

DISCUSSION

The recursive formula can, in general, be found when the numerator and denominator in the expression for $b^{t+1}_{ij_i}$ are homogeneous functions with the same degree.

The biological system converges to the equivalent equilibrium point by the opposing processes of differential biomass increase due to natural fecundity and the limiting effect of the environment. This selection process under competition is a general phenomenom and our simplified model represents a special case. It will be of particular interest to analyse the rates of convergence, and the effects of stochastic factors such as genetic drift or founder populations.

This iterative approach to the solution of a game is an example of an emerging and highly important part of game theory. Owen (1968) refers to the general process as solution by "fictitious" play. Other examples may be found in Rapaport (1970), particularly the work of Brown and Von Neumann(1950), Brown (1951), Robinson (1951), and Bellman (1953).

ACKNOWLEDGEMENTS

We thank David Gibo, Michael Gates, Dennis Lynn, for their helpful discussion.

BIBLIOGRAPHY

Bellman, R. 1953. On a new iterative algorithm for finding the solutions of games linear programming problems. Research Memo. Rand Corp. 473 p.

Brown, G.W. 1951. Iterative solutions of games by fictitious play. In Activity Analysis of Productivity and Allocation. Koopmans Ed. Wiley. 377-380.

Brown, G.W. and J. Von Neumann. 1950. Solutions of games by differential equations. Contribution to the theory of games. Vol. I. Princeton U. Press.

Fisher, R.A. 1930. The Genetical Theory of Natural Selection. Clarendon, Oxford, 272 p.

Karlin, S. and M.W. Feldman. 1970. Convergence to Equilibrium of the two locus additive viability model. J. Appl. Prob. 7(2): 262-271.

Lewontin, R.C. 1961. Evolution and the theory of games. J. Theoret. Biol. 1: 382-403.

Marchi, E. and R.I.C. Hansell. 1973a. Generalization on the Parsimony Question in Evolution. Math. Biosci. 17: 11-34.

Marchi, E. and R.I.C. Hansell. 1973b. A Framework for Systematic Zoological Studies with Game Theory. Math. Biosci. 16: 31-58.

Marchi, E. and R.I.C. Hansell. 1974. Fuzzy Aspects of the Parsimony Problem in Evolution. Math. Biosci. (in press).

Owen, G. 1968. Game Theory. W.B. Saunders Co., 228 p.

Rapoport, A. 1970. N-Person Game Theory. Univ. Michigan Press. 329 p.

Robinson, J. 1951. An Iterative method of solving a game. Annals of Mathematics, 54: 296-301.

Slobodkin, L. 1967. Existential game. In Population Biology and Evolution. R.C. Lewontin Ed., Syracuse U. Press.

Wright, S. 1931. Evolution in Mendelian Populations. Genetics. 16: 97-159.

COEXISTENCE OF SPECIES IN A DISCRETE SYSTEM

U. G. HAUSSMANN

Department of Mathematics
The University of British Columbia
Vancouver 8, B.C.

1. Introduction

The principle of competitive exclusion states that n species competing for fewer than n resources cannot coexist indefinitely. Using his competition model, Volterra (1927) proved this result mathematically for the case of one resource. Experimental verification was supplied by the work of Gause (1934) with *Paramecium caudatum* and *Paramecium aurelia*, as well as by Park's work with the flour beetle. Later MacArthur and Levins (1964) argued that the result extends to more than one resource. Independently Rescigno and Richardson (1965) gave a mathematical proof of the exclusion principle for a model more general than Volterra's, allowing any number of resources, or as they call them, niches. Levin (1970) simplified their proof a little and observed that the niches did not have to be resources - he called them limiting factors. From this the general form of the principle follows: *If n species coexist, then they must be limited by at least n independent limiting factors.*

However all the above analysis is based on a differential equations model which is *linear* in the limiting factors or resources. Haussmann (1973) removed this restriction. In the present work we extend this result to a difference equation model, and we also show that small time dependent perturbations of the model do not destroy coexistence as defined below.

2. Coexistence in the Model

Consider a community of n components with respective densities at time t: $x_1^t, x_2^t, \ldots, x_n^t$. Usually the components are species, but they could be segments or groups of species or even concentrations of nutrients etc. Let $y_1^t, \ldots, y_m^t$ be parameters determined by the environment, hence affecting the species, but which themselves are independent of the latter, although possibly time varying, for example climatic variables.

Assume the dynamics of the community are given by

$$(1) \qquad x_i^{t+1} = x_i^t \exp g_i(x^t, y^t) , \quad i = 1, 2, \ldots, n ; \quad t = 0, 1, 2, \ldots ,$$

where $x^t = \mathrm{col}(x_1^t, x_2^t, \ldots, x_n^t)$, $y^t = \mathrm{col}(y_1^t, y_2^t, \ldots, y_m^t)$. For each initial condition x^0 , (1) has a solution - a sequence of n-tuples - called an <u>orbit</u>.

Assume for now that y is constant, so we suppress it in the notation. The community is said to <u>coexist indefinitely</u> if the orbit describing the densities is a steady-state, i.e. contains a finite number of points, say N , for which none of the densities are zero. Moreover the steady-state must be exponentially asymptotically stable, i.e. if the solution is slightly perturbed away from the steady-state then it tends to return to it at an exponential (or geometric) rate. Observe of course that steady states which are not stable would not persist.

Of course the y's are, in general, not constant, and so no steady-state would exist, but we hope that they vary slowly measured on the time scale of the x's and so may be taken as constant. The idea is that the system will approach the steady-state relatively quickly (i.e. at an exponential rate) and then will vary slowly as the "steady-state" moves because of the time dependence of the y's . That this hope is justified follows from the first theorem (see appendix for proofs).

<u>Theorem 1</u>. If $\{x^t\}$ is an exponentially asymptotically stable steady-state of

$$(2) \qquad x_i^{t+1} = f_i(x^t) \qquad\qquad i = 1, 2, \ldots, n ,$$

and if y^t is a solution of

$$(3) \qquad y_i^{t+1} = f_i(y^t) + \varepsilon\, h_i^t(y^t)$$

with ε small and y^0 close to x^0, then (i) y^t remains close to x^t, and (ii) y^t is exponentially asymptotically stable.

3. The Exclusion Principle

On the basis of the previous result we shall restrict our attention to systems where y is constant, i.e.

$$(4) \qquad x_i^{t+1} = x_i^t \exp g_i(x^t) , \qquad i = 1, 2, \ldots, n .$$

Frequently in ecology $g_i(x)$ depends not on the x_j's individually but rather on some combinations $z_1, z_2, \ldots, z_p$, where, of course, each z_k depends on x. We call these functions $z_1, \ldots, z_p$ the limiting factors of the community. Their number and form may depend on the value of x.

Consider now a steady-state $\{x^1, x^2, \ldots, x^N\}$. Then $\{z_1, \ldots, z_p\}$ is a minimal independent set of limiting factors of this orbit if each $g_i(x)$ can be written as

$$g_i(x) = \phi_i[z(x)] , \qquad i = 1, 2, \ldots, n ,$$

for x in a neighbourhood of x^t, $t = 1, 2, \ldots, N$, where $z(x) = \mathrm{col}(z_1(x), \ldots, z_p(x))$, and if, moreover, no smaller set of limiting factors has this property. For further discussion of limiting factors see Haussmann (1973). The effect of z_j at z^0 is the n-tuple $(\partial\phi/\partial z_j)(z^0)$ whose ith component is $(\partial\phi_i/\partial z_j)(z^0)$. These effects determine a subspace S of n-dimensional space, the smallest subspace containing $\{(\partial\phi/\partial z_j)(z(x^t)) : j = 1, 2, \ldots, p ; t = 1, 2, \ldots, N\}$. The dimension of S is called the index of the limiting factors along the orbit, and is denoted by J. Hence J is the largest number of independent effects which the limiting factors can have along the orbit.

Usually J is difficult to compute. However it can be shown, Haussmann (1973), (i) if the orbit corresponds to a point, i.e. $N = 1$, then $J \leqq p$, and (ii) if each ϕ_i is linear in z, i.e.

$$\phi_i(z) = a_{i1} z_1 + a_{i2} z_2 + \dots + a_{ip} z_p + b_i ,$$

as assumed by all the other researchers in the field, then $J \leqq p$.

As an example consider the following predator-prey system, similar to one studied by Haussmann (1971).

$$x_1^{t+1} - x_1^t = x_1^t \left\{ \frac{a_1}{1 + b_1 x_1^t} - d_1 - \frac{a_2 x_2^t}{1 + b_2 x_1^t} \right\}$$

$$x_2^{t+1} - x_2^t = x_2^t \left\{ \frac{e_2 a_2 x_1^t}{1 + b_2 x_1^t} - d_2 \right\}$$

where x_1 is the prey density and x_2 is the predator density. In the form (4)

$$g_1(x) = \ln \left\{ 1 + \frac{a_1}{1 + b_1 x_1} - d_1 - \frac{a_2 x_2}{1 + b_2 x_1} \right\}$$

$$g_2(x) = \ln \left\{ 1 + \frac{e_2 a_2 x_1}{1 + b_2 x_1} - d_2 \right\} .$$

Three obvious limiting factors are the specific energy intake of the prey, $a_1/(1 + b_1 x_1)$, the specific loss to predation, $a_2 x_2/(1 + b_2 x_1)$, and the specific energy intake of the predator, $e_2 a_2 x_1/(1 + b_2 x_1)$. Here $e_2 \leq 1$ is an efficiency coefficient and d_1, d_2 are natural mortality rates. Then $\phi_1 = \ln(1 - d_1 + z_1 - z_2)$, $\phi_2 = \ln(1 - d_2 + z_3)$ and the ϕ_i's are not linear in z_1, z_2, z_3 , and, moreover, the latter are not minimal independent. Such a set would be $z_1 = 1/(1 + b_1 x_1)$, $z_2 = x_2/(1 + b_2 x_1)$ with index 2.

We can now give the main result.

Theorem 2. (Exclusion Principle) If a community of n components coexists indefinitely then the index is n .

Observe that there may be fewer than n limiting factors, but the community can still persist in an oscillating manner if the limiting factors have different effects over different segments of the oscillations.

The experiments of Paine (1966) are interesting in this connection. He found that the removal of the top predator (Pisaster ochraceus) in a simple foodweb led to the elimination of several of the prey species. An explanation based on our work is that several effects of the limiting factors, when applied to the whole community including Pisaster, are independent; whereas these effects, when restricted to the community without Pisaster, are no longer independent. The removal of Pisaster leaves the resulting community too large for stability according to the extended principle, so that several species must disappear before a steady-state can be re-established.

4. Appendix

If x^t is a solution of (2) it is defined to be exponentially asymptotically stable if there is $q < 1$ such that for all $\nu > 0$ there is $\delta > 0$ if y^t is another solution of (2) with $|y^s - x^s| < \delta$, then $|y^t - x^t| < \nu q^{t-s}$ and $\liminf_{\nu \to 0} \nu/\delta < \infty$.

Consider the equation

$$(5) \qquad w^{t+1} = A_t w^t + u_t$$

and set $\Phi_{ts} = A_{t-1} A_{t-2} \dots A_s$, $\Phi_{tt} = I$, the $n \times n$ identity matrix. Then

$$(6) \qquad w^t = \Phi_{ts} w^s + \sum_{r=s+1}^{t} \Phi_{tr} u_{r-1}, \quad t \geqq s .$$

Lemma 1. Assume that for any $\eta > 0$ there is $r_0 > 0$ such that $|u_t| < \eta |w^t|$ if $|w^t| < r_0$. If the zero solution of (5) is exp. as. stable, then $|\Phi_{ts}| \leqq K \rho^{t-s}$ for some $\rho < 1$, $K \geqq 1$.

Proof: If $|w^s| < \delta$ with $\nu = r_0$, it follows from (6) that

$$|\Phi_{ts} w^s| \leqq \nu q^{t-s} + \nu \eta \sum_{r=s+1}^{t} |\Phi_{tr}| q^{r-s-1} ,$$

so $$|\Phi_{ts}|\, q^s \leqq q^t \nu/\delta + \nu\eta/(\delta q) \sum_{r=s+1}^{t} |\Phi_{tr}|\, q^r \,, \quad s = 0, 1, \ldots, t \,.$$

An argument similar to that used by Miller (1968), page 22, applied to the variable s yields

$$\begin{aligned} |\Phi_{ts}| &\leqq \nu \{q \exp \nu\eta/(\delta q)\}^{t-s}/\delta \\ &\leqq K\, q[\exp(K\eta/q)]^{t-s} \end{aligned}$$

where $\nu/\delta \leqq K$ since $\liminf \nu/\delta < \infty$. By choosing r_0 sufficiently small, $\rho = q \exp(K\eta/q)$ can be made less than 1.

<u>Corollary 1</u>. Under the conditions of lemma 1 the zero solution of the linear system

$$w^{t+1} = A_t w^t$$

is also exp. as. stable.

<u>Proof of Theorem 1</u>. We assume that the orbit C is given by $\{x^1, x^2, \ldots, x^N\}$. For the proof we require:

(I) there is a $D_o > 0$ such that each f_i is continuously differentiable in a ball of radius D_o about each x in C;

(II) there is a constant k_1 such that $|h^t(x)| \leqq k_1$ for any t and each x in C;

(III) there is a constant k_2 such that

$$|h^t(x + \xi) - h^t(x + \xi')| \leqq k_2|\xi - \xi'|$$

for any x in C if $|\xi|$, $|\xi'| < D_o$.

Let $\bar{x}^t$ be another solution of (2) and put $\zeta^t = \bar{x}^t - x^t$. Then

$$\zeta^{t+1} = f'(x^t)\, \zeta^t + \chi_t(\zeta^t)$$

where $|\chi_t(\zeta)| \leqq v(r_1)\, |\zeta|$ if $|\zeta| < r_1 < D_o$, and

$$v(r_1) = \sup\{\int_0^1 [f'(x + \bar{\zeta} + \tau(\zeta - \bar{\zeta})) - f'(x)]d\tau : x \text{ in } C , |\zeta| , |\bar{\zeta}| \leqq r_1\} ,$$

so that $v(r_1)$ decreases monotonically to 0 as $r_1 \to 0$ because of (I). Hence with $A^t = f'(x^t)$, lemma 1 implies $|\Phi_{ts}| \leqq K \rho^{t-s}$, $K \geqq 1$.

Now set $w^t = y^t - x^t$. Then

$$(7) \qquad w^{t+1} = A_t s^t + F_t(w^t, \varepsilon) ,$$

where $F_t(w, \varepsilon) = \chi_t(w) + \varepsilon h^t(x^t + w)$. To prove (i) we shall find w^t as a fixed point of a contraction mapping - this will give a bound on w^t. Q is in $\ell_\infty^+(R^n)$ if $Q = \{Q^0, Q^1, \ldots\}$, $Q^t \in R^n$, and $\sup\{|Q^t|: t \geqq 0\} = ||Q||_\infty < \infty$. For fixed w^0 and Q in $\ell_\infty^+(R^n)$ let $w[Q]$ denote the unique solution of

$$w^{t+1} = A_t w^t + Q^t .$$

Let $S^n(D, w^o) = \{\xi \text{ in } \ell_\infty^+(R) : ||\xi||_\infty \leqq D , \xi^0 = w^0\}$, and consider a function $Q : S^n(D, w^o) \to S^n(\beta(D), w^o)$ such that $||Q(\xi) - Q(\xi')||_\infty \leqq \alpha(D) ||\xi - \xi'||_\infty$ if ξ, ξ' are in $S^n(D, w^o)$. Let $T(\xi) = w[Q(\xi)]$. Using (6) it follows that $T : S^n(D, w^o) \to S^n(D, w^o)$ if

(a) $K|w^o| + K \beta(D)/(1 - \zeta) \leqq D$,

and that T is a contraction if

(b) $K \alpha(D) < 1 - \zeta$.

If $Q(\xi)^t = F_t(\xi^t, \varepsilon)$, then the fixed point of T is the solution w^t of (7) and so $|w^t| \leqq D$. It can be shown using (II) and (III) that $\alpha(D)$ can be taken as $\varepsilon k_2 + v(D)$ and $\beta(D)$ as $\varepsilon k_1 + [\varepsilon k_2 + v(D)]D$ if $D < D_0$. If $|w^o| < D/(3K)$, $K \varepsilon k_1 \leqq (1 - \rho)D/3$, and $K[\varepsilon k_2 + v(D)] \leqq (1 - \rho)/3$, then (a) and (b) are satisfied. Set $D_1(\varepsilon) = 3K k_1\varepsilon/(1 - \rho)$, $D_2(\varepsilon) = v^{-1}[(1-\rho)/(3K) - \varepsilon k_2]$. Then (a) and (b) hold for any D such that $D_1(\varepsilon) \leqq D \leqq D_2(\varepsilon)$. There exists

$\varepsilon_1 > 0$ such that

$$v[3K\ k_1\ \varepsilon_1/(1-\rho)] = (1-\rho)/(3K) - \varepsilon_1\ k_2 ,$$

from which it follows that $D_1(\varepsilon) \leqq D_2(\varepsilon)$ if $\varepsilon \leqq \varepsilon_1$. Let $\varepsilon_o = \max\{\varepsilon: \varepsilon \leqq \varepsilon_1 , D_1(\varepsilon) \leqq D_0\}$. Hence for $\varepsilon < \varepsilon_0$ we set $D = D_1(\varepsilon)$ and conclude that if $|w^0| \leqq k_1\varepsilon/(1-\rho)$, then $|w^t| \leqq 3K\ k_1\varepsilon/(1-\rho)$, and (i) is established.

To prove (ii) we now let $y^t + w^t$ be another solution of (3). Then

$$w^{t+1} = A_t\ w^t + \psi_t(w^t) + \varepsilon[h^t(y^t + w^t) - h^t(y^t)]$$

where $|\psi_t(w)| \leqq v(r_o)\ |w|$ if $|w| < r_o/2$ and if $|y^t - x^t| < r_o/2$, i.e. if $\varepsilon < r_0(1-\rho)/(6K\ k_1)$. Hence if $|w^s| < r_0/2 \leqq D_o/2$ for $s < t$, then from (6) and (III) it follows that

$$|w^t|\rho^{-t} \leqq K|w^o| + \sum_{r=1}^{t} K[v(r_o) + \varepsilon\ k_2]\ |w^{r-1}|\ \rho^{-r}$$

and so from Miller (1968), p. 21,

$$|w^t| \leqq K|w^o|\ \rho^t \exp\{tK[v(r_o) + \varepsilon\ k_2]\} .$$

Now choose r_0 and ε so small that $\rho \exp K[v(r_o) + \varepsilon k_2] \leqq \rho_o < 1$. Then $|w^t| \leqq K|w^o|\ \rho_o^t < r_o/2$ if $|w^o| < r_o/(2K)$. This proves the theorem.

<u>Proof of Theorem 2</u>. We assume that the steady-state orbit is $C = \{x^1, x^2, \ldots, x^N\}$, and that each g_i is continuously differentiable in a D_o-ball about any x in C. Assume $x^t + \xi^t$ is another solution of (4). Then

$$\text{(8)} \qquad \xi_i^{t+1} = \exp g_i(x^t)\ \{\xi_i^t + x_i^t \sum_{j=1}^{n} (\partial g_i/\partial x_j)(x^t)\xi_j^t\} + \psi_i^t(\xi^t)\}$$

where for any $\mu > 0$ there is a $\delta > 0$, such that if $|\xi| < \delta$, then $|\psi_i^t(\xi)| < \mu|\xi|$.

Let η^t satisfy

$$\eta_i^{t+1} = \eta_i^t + \sum_{j=1}^{n} (\partial g_i/\partial x_j)(x^t)\, x_j^t\, \eta_j^t \, . \tag{9}$$

If $\zeta_i^t = x_i^t \eta_i^t$, then

$$\zeta_i^{t+1} = \exp g_i(x^t)\, \{\zeta_i^t + x_i^t \sum_{j=1}^{n} (\partial g_i/\partial x_j)\, (x^t)\, \zeta_j^t\} \, . \tag{10}$$

If the community coexists then the zero solution of (8) is exp. as. stable and so by corollary 1 the zero solution of (10) is also exp. as. stable. We will now show that this is not the case, in fact $|\zeta^t| \not\to 0$ as $t \to \infty$ for suitable ζ^o.

Recall $g_i(x) = \phi_i[z(x)]$ where $(z_1, \ldots, z_p)$ are the limiting factors. If the index is less than n then there is an n-tuple $\beta \neq 0$ such that

$\beta \cdot (\partial\phi/\partial z^j)(z(x)) = 0$ for all x in C and $j = 1, 2, \ldots, p$. Hence from (9) $\beta \cdot \eta^{t+1} = \beta \cdot \eta^t$ for all t, or $\beta \cdot \eta^t = c$ where c is a constant, $c \neq 0$ for suitable η^o. It follows that $\sum_{i=1}^{n} (\beta_i/x_i^t)\, \zeta_i^{t+kN} = c \neq 0$ for any $t = 1, 2, \ldots, N$, and any $k = 0, 1, 2, \ldots$, so that $|\zeta^t| \not\to 0$ as $t \to \infty$.

References

Gause, G.F. The Struggle for Existence, Williams and Wilkins, Baltimore, Md. (1934).

Haussmann, U.G. Math Biosciences 11, 291-316 (1971).

_____________. Theoret. Pop. Biology 4, 31-41 (1973).

Levin, S.A. Amer. Naturalist 104, 413-424 (1970).

MacArthur, R. and R. Levins. Nat. Acad. Sci. Proc. 51, 1207-1210 (1964).

Miller, K.S. Linear Difference Equations, Benjamin, New York, (1968).

Paine, R.T. Amer. Naturalist 100, 65-76 (1966).

Rescigno, A. and I.W. Richardson. Bull. Math. Bioph. 27, 85-89 (1965).

Volterra, V. Variazioni e fluttuazioni del numero d'individui in specie animali conviventi, R. Comit. Talass. Italiano, Memoria 131 (1927), Venezia.

ASYMPTOTIC BEHAVIOR AND STABILITY IN EPIDEMIC MODELS

Herbert W. Hethcote

Department of Mathematics
The University of Iowa
Iowa City, Iowa 52242

Abstract. Deterministic epidemic models of SIS and SIR type are considered where births and deaths occur at equal rates with all newborns being susceptible. In an SIS model the effect of births and deaths is to raise the threshold which determines the number of infectives as time approaches infinity. The asymptotic behavior in an SIS model is also indicated when the contact rate is periodic and when carriers are present. Asymptotic stability regions are determined for the equilibrium points in a host and vector SIS model and in an SIR epidemic model.

THE SIS EPIDEMIC MODEL

In an SIS model a fixed population size N is divided into a class of susceptibles $S(t)$ and a disjoint class of infectives $I(t)$ which vary with time t such that $I(0) = I_0 > 0$ and $S(t) + I(t) = N$. The rate of increase of infectives due to new infections is given by $\beta I(t)S(t)$ where β is called the infection or contact rate. We assume that infection does not give immunity and let the rate at which infectives recover and become susceptible again be given by $\gamma I(t)$ where γ is called the recovery rate. We also assume that births and deaths occur within the fixed population size N at equal rates and that all newborns are susceptible. If δ is the specific birth and death rate, then the rate of decrease of infectives is given by δI. Thus the differential equation for $I(t)$ is

$$I'(t) = \beta I(t)(N - I(t)) - \gamma I(t) - \delta I(t). \tag{1}$$

The number of susceptibles can always be found from $I(t)$ by using $S(t) = N - I(t)$. Other epidemic models are described in the monograph of Bailey [2].

The differential equation (1) is similar to the differential equation for $I(t)$ in the SIS model with no birth and death which we considered in [4] except that we now have $\gamma + \delta$ in place of γ. Since (1) is a Riccati differential equation, we can use the method outlined in [4] to show that the unique solution of the initial value problem is

$$I(t) = \begin{cases} \dfrac{\exp(kt)}{\beta(e^{kt}-1)/k + 1/I_0} & k \neq 0 \\ \dfrac{1}{\beta t + 1/I_0} & k = 0 \end{cases}$$

where we now have $k = N\beta - \gamma - \delta$. Only the asymptotic representation was given in [4]. Here we give the complete asymptotic expansion so that the manner in which the asymptotic limit is approached is apparent and approximate solutions are easily determined for large t. By using

$$\frac{1}{1-x} = \sum_{n=0}^{m} x^n + \frac{x^{m+1}}{1-x}, \qquad |x| < 1$$

we find that the asymptotic expansions are

$$I(t) \sim \frac{-kI_0 \exp(kt)}{\beta I_0 - k} \sum_{n=0}^{\infty} \left[\frac{\beta I_0 \exp(kt)}{\beta I_0 - k}\right]^n, \qquad k < 0$$

$$I(t) \sim \frac{1}{\beta t} \sum_{n=0}^{\infty} (-1)^n (I_0 \beta t)^{-n}, \qquad k = 0$$

$$I(t) \sim \left(N - \frac{\gamma+\delta}{\beta}\right) \sum_{n=0}^{\infty} \left[\frac{(\beta I_0 - k)\exp(-kt)}{\beta I_0}\right]^n, \qquad k > 0.$$

Consequently, $I(t) \longrightarrow N - (\gamma+\delta)/\beta$ as $t \longrightarrow \infty$ for $N > (\gamma+\delta)/\beta$ and $I(t) \longrightarrow 0$ as $t \longrightarrow \infty$ for $N \leq (\gamma+\delta)/\beta$. Hence the threshold value of N, which determines the asymptotic behavior of $I(t)$, is now $(\gamma+\delta)/\beta$. This result with $\delta = 0$ was obtained by Weiss and Dishon [6]. The effect of births and deaths at a rate δ on the SIS model is that the threshold is raised by δ/β. Although we omit the details, it can be shown that births and deaths at rate δ would also raise the threshold if $\beta(t)$ were periodic [4]. As noted in [4], if the total population grows exponentially or linearly, then $I(t) \longrightarrow \infty$ as $t \longrightarrow \infty$. If there is no recovery $(\gamma = 0)$, then our model is of SI type with births and deaths and the threshold is given by δ/β. If both $\gamma = 0$ and $\delta = 0$, then our model is the simple SI epidemic model of Bailey [1; 2, p. 20] and $I(t)$ always approaches N as $t \longrightarrow \infty$.

Since we will be considering the stability of the equilibrium points for a SIR model, we use the asymptotic expansions to determine the following stability results for the above SIS model. The critical point $I = 0$ is asymptotically stable for $N \le (\gamma+\delta)/\beta$ and unstable for $N > (\gamma+\delta)/\beta$. The critical point $I = N - (\gamma+\delta)/\beta$ is not in the region $0 \le I \le N$ for $N < (\gamma+\delta)/\beta$ and is asymptotically stable for $N \ge (\gamma+\delta)/\beta$.

Epidemic models of SIS type with carriers were considered in [4]. If births and deaths occur within the fixed population size N at rate δ and all newborns are susceptible, then the results are similar to those in [4] except that $\gamma + \delta$ is used in place of γ in all expressions. The net effect is to shift the asymptotic limit when the limit is a positive constant and to change the rate at which $I(t) \longrightarrow 0$ otherwise. Details are omitted and can be supplied by the reader.

THE HOST AND VECTOR SIS MODEL

We now consider a host and vector SIS epidemic model where the notation is consistent with the earlier SIS model. In the differential equations

$$
(2) \qquad
\begin{aligned}
I_1'(t) &= \beta_1 I_2(N_1 - I_1) - (\gamma_1+\delta_1)I_1, \\
I_2'(t) &= \beta_2 I_1(N_2 - I_2) - (\gamma_2+\delta_2)I_2,
\end{aligned}
$$

γ_1, δ_1, γ_2, and δ_2 are nonnegative; and β_1, β_2, and $\gamma_1+\delta_1+\gamma_2+\delta_2$ are positive. We have let $I_1+S_1=N_1$ and $I_2+S_2=N_2$. We now analyse the asymptotic behavior of the two dimensional autonomous system (2) in the region $D: 0 \le I_1 \le N_1$, $0 \le I_2 \le N_2$. For $N_1N_2-\rho_1\rho_2 \neq 0$, the critical points are the origin and

$$
(3) \qquad \left(\frac{N_1N_2-\rho_1\rho_2}{N_2+\rho_1}, \frac{N_1N_2-\rho_1\rho_2}{N_1+\rho_2}\right)
$$

where $\rho_1=(\gamma_1+\delta_1)/\beta_1$ and $\rho_2=(\gamma_2+\delta_2)/\beta_2$.

The systems of differential equations which we consider are perturbed linear systems of the form

$$
(4) \qquad
\begin{aligned}
I_1'(t) &= a\,I_1 + b\,I_2 + f_1(I_1,I_2), \\
I_2'(t) &= c\,I_1 + d\,I_2 + f_2(I_1,I_2).
\end{aligned}
$$

The roots of the characteristic equation of the linear part of (4) are

$$
(5) \qquad r\pm = \tfrac{1}{2}\left[a+d\pm\sqrt{(a+d)^2-4(ad-bc)}\right] = \tfrac{1}{2}\left[a+d\pm\sqrt{(a-d)^2+4bc}\right].
$$

When $a+d<0$ and $bc \ge 0$, both roots are negative if $ad-bc>0$ and one root is positive if $ad-bc<0$.

We first consider the critical point at the origin. The system (2) is almost linear since the perturbation terms are quadratic and hence tend to zero faster than the linear terms as the origin is approached [3]. Comparing (2) with (4) we find that $a+d<0$, $bc>0$, and $ad-bc=-\beta_1\beta_2[N_1N_2-\rho_1\rho_2]$ so that the origin is asymptotically stable (improper node) for (2) if $N_1N_2-\rho_1\rho_2<0$ and unstable (saddle point) if $N_1N_2-\rho_1\rho_2>0$. For $N_1N_2-\rho_1\rho_2<0$, the origin is the only critical point in D. Since $I_1 = 0$ implies $I_1'(t) = \beta_1 N_1 I_2 \geq 0$ and $I_1 = N_1$ implies $I_1'(t) = -(\gamma_1+\delta_1)I_1 \leq 0$, no trajectory in D leaves through the left or right side. Similarly no trajectory leaves D through the top or bottom. Since limit cycles must contain a critical point, there are no limit cycles in D. Thus every trajectory in D must approach the origin and, consequently, D is a region of asymptotic stability for the origin if $N_1N_2 - \rho_1\rho_2 < 0$. Moreover, all trajectories in D enter the origin tangent to a line with slope

$$M = \left[-(\gamma_1+\delta_1) + (\gamma_2+\delta_2) + \sqrt{[(\gamma_1+\delta_1)-(\gamma_2+\delta_2)]^2 + 4\beta_1\beta_2N_1N_2}\right]\Big/2\beta_1N_1.$$

We now consider the critical point (3) which lies in D if $N_1N_2 - \rho_1\rho_2 > 0$. We translate the critical point (3) to the origin by letting

$$U = I_1 - \frac{N_1N_2 - \rho_1\rho_2}{N_2 + \rho_1}, \quad V = I_2 - \frac{N_1N_2 - \rho_1\rho_2}{N_1 + \rho_2}.$$

The differential equations (2) become

$$(6) \qquad \begin{aligned} U'(t) &= \beta_1\left[-UV - N_1\frac{N_2+\rho_1}{N_1+\rho_2}U + \rho_1\frac{N_1+\rho_2}{N_2+\rho_1}V\right], \\ V'(t) &= \beta_2\left[-UV + \rho_2\frac{N_2+\rho_1}{N_1+\rho_2}U - N_2\frac{N_1+\rho_2}{N_2+\rho_1}V\right]. \end{aligned}$$

This system is also almost linear and we again use the notation and results following (4) and (5). By comparing (6) with (4), we find that $a + d < 0$, $bc \geq 0$, and $ad - bc = \beta_1\beta_2(N_1N_2 - \rho_1\rho_2)$. Thus the critical point (3) is asymptotically stable for (2) if $N_1N_2 - \rho_1\rho_2 > 0$.

For the unstable saddle point at the origin when $N_1N_2 - \rho_1\rho_2 > 0$, the only solutions which approach the origin lie on a line [3]. This line has slope $m = (r_- - a)/b$ which is negative so that no trajectory in D approaches the origin. For $N_1N_2 - \rho_1\rho_2 > 0$, the same arguments used earlier show that no trajectory leaves D at a boundary point. We now show that there are no periodic solutions contained entirely in D. Assume $I_1 = \varphi(t), I_2 = \psi(t)$ is a periodic solution denoted by C with interior region R lying entirely in D. Using Green's theorem on the right hand sides of (2) which we denote by F_1 and F_2, we arrive at the following contradiction.

$$0 = \int_C [F_1(\varphi,\psi)d\psi - F_2(\varphi,\psi)d\varphi] = \iint_R (F_{1\varphi}+F_{2\psi})dA$$

$$= \iint_R [-\beta_1\psi - (\gamma_1+\delta_1) - \beta_2\varphi - (\gamma_2+\delta_2)]dA < 0.$$

Thus every trajectory in D, except the fixed point at the origin, must approach the critical point (3) and $D - \{(0,0)\}$ is a region of asymptotic stability for the critical point (3) if $N_1N_2 - \rho_1\rho_2 > 0$. Moreover, the critical point (3) is an improper node if the discriminant is positive and is a node or spiral if the discriminant is zero. We omit the details of how the trajectories approach the improper node.

If $N_1N_2 - \rho_1\rho_2 = 0$, the critical points of the linear part of (2) are on the line $I_2 = \rho_1 I_1/N_1$ and are not isolated so the usual methods do not apply. In the region $A = D \cap \{(I_1, I_2) : I_2 \le \rho_1 I_1/N_1\}$, we have $I_1'(t) \le 0$ with equality only at the origin so that trajectories must approach the origin or cross into $D - A$. In $D - A$, $I_2'(t) \le 0$ with equality only at the origin so that trajectories must approach the origin or cross into A. Thus D is also a region of asymptotic stability for the origin when $N_1N_2 - \rho_1\rho_2 = 0$.

THE SIR EPIDEMIC MODEL

In an SIR model, a community of fixed size N is divided into the classes of susceptibles $S(t)$, infectives $I(t)$, and removed individuals $R(t)$ who are isolated, dead, or recovered and immune. Here we also assume that births and deaths occur at equal rates and that all newborns are susceptible. Using the same notation as in the SIS models, the differential equations are

$$(7)\qquad \begin{aligned} S'(t) &= -\beta IS + \delta N - \delta S \\ I'(t) &= \beta IS - \gamma I - \delta I \end{aligned}$$

where β and δ are positive and γ is nonnegative. The number of removed individuals can be found at any time t by using $R(t) = N - I(t) - S(t)$.

We now analyse the asymptotic behavior of the system (7) in the region $D: S > 0,\ I > 0,\ S + I < N$. Since $S = 0$ implies $S'(t) = \delta N > 0$, $I = 0$ is a trajectory, and $S + I = N$ implies $S'(t) + I'(t) = -\gamma I \le 0$, no trajectory leaves D. The critical points of the system (7) are

$$(8)\qquad S = N, \quad I = 0$$

and

$$(9)\qquad S = \rho, \quad I = \delta(N-\rho)/(\gamma+\delta)$$

where $\rho = (\gamma+\delta)/\beta$. If we translate the critical point (8) to the origin by letting $S = N(1+U)$, then the system (7) becomes

$$(10)\qquad \begin{aligned} U'(t) &= -\delta U - \beta I - \beta IU, \\ I'(t) &= [\beta N - \gamma - \delta]I + \beta NIU. \end{aligned}$$

This is an autonomous perturbed linear system with quadratic perturbation terms and the roots of the characteristic equation of the linear part are $-\delta$ and $\beta N - \gamma - \delta$. If $N < \rho$, both roots are negative and the critical point (8) is asymptotically stable (improper node). For $N < \rho$, (8) is the only critical point in $\overline{D}$ (the closure of D) and there are no limit cycles in D since (8) is on the boundary of D; consequently, $\overline{D}$ is a region of asymptotic stability for the equilibrium point (8) if $N < \rho$. If $N > \rho$, one root is positive and the critical point (8) is unstable (saddle point).

If we translate the critical point (9) to the origin by letting $S = \rho(1+U)$, $I = \delta(N-\rho)(1+V)/(\gamma+\delta)$, then the system (7) becomes

$$\rho U'(t) = -\delta NU - \delta(N-\rho)V - \delta(N-\rho)UV,$$
(11)
$$V'(t) = (\gamma+\delta)U + (\gamma+\delta)UV.$$

The roots of the characteristic equation of the linear part of this perturbed linear system are

(12)
$$r_{\pm} = \left[-\delta N \pm \sqrt{\delta^2 N^2 - 4\rho\delta(\gamma+\delta)(N-\rho)}\right] \Big/ 2\rho .$$

If $N < \rho$, one root is positive and the critical point (9) is an unstable saddle point which lies outside of $\overline{D}$. If $N > \rho$, both roots have negative real part and the critical point (9) is asymptotically stable and lies in D. Moreover, the critical point (9) is an improper node if the discriminant in (12) is positive and a spiral if the discriminant is negative.

We analyse the region of asymptotic stability for the critical point (9) by using Liapunov's direct method [3, p. 214]. The function

$$L(U,V) = \rho[U - \ln(1+U)]/\delta + (N-\rho)[V - \ln(1+V)]/(\gamma+\delta)$$

is a positive definite scalar function defined on the set

$$\Omega = \{(U,V) : U > -1,\ V > -1,\ \rho(1+U)/\delta + (N-\rho)(1+V)/(\gamma+\delta) < N\},$$

which is the set D in the UV coordinate system. The function $L(U,V)$ is continuously differentiable on Ω and $dL/dt = -NU^2/(1+U) \leq 0$ in Ω. For some values of $\lambda \geq 0$, the sets $C_\lambda = \{(U,V) : L(U,V) \leq \lambda\}$ are closed bounded subsets of Ω and contain the origin. Since $U = 0$ implies $U'(t) = -\delta(N-\rho)V/\rho$ which is not zero except when $V = 0$, the origin is the only invariant subset of $E = \Omega \cap \{(U,V) : dL/dt = 0\}$. Thus every solution starting in C_λ approaches the origin as $t \to \infty$. We let $\hat{\lambda}$ be the largest λ such that C_λ is contained in Ω. The set $C_{\hat{\lambda}}$ touches the boundary of Ω at the intersection of the lines $U = V$ and $\rho(1+U)/\delta + (N-\rho)(1+V)/(\gamma+\delta) = N$.

Returning to the SI plane, we now have a region of asymptotic stability for the critical point (9) which touches the boundary of D. Since any limit cycle must contain a critical point, there can be no limit cycles in D. Thus $\bar{D} - \{(S,0) : 0 \leq S \leq N\}$ is a region of asymptotic stability for the equilibrium point (9).

If $N = \rho$, the critical points of (7) lie on the line $I = \delta(1-S/N)/\beta$ and are not isolated so that the previous methods do not apply. However, from (7) we see that $N = \rho$ implies $I'(t) \leq 0$ with equality in $\bar{D}$ only if $I = 0$. Thus all trajectories approach the $I = 0$ line, but since $I = 0$ is a trajectory, all trajectories must approach the critical point (8). Hence $\bar{D}$ is also a region of asymptotic stability for the critical point (8) if $N = \rho$.

The assumption that δ is positive in (7) is crucial to the previous stability results. If $\delta = 0$, then all points on the line $I = 0$ are critical points. By analysing the trajectories in the first quadrant as in Hethcote and Waltman [5], one can show that the critical points $I = 0$, $0 \leq S \leq \rho$ are stable, but not asymptotically stable; and the critical points $I = 0$, $\rho < S < \infty$ are unstable.

The local asymptotic behavior of an SIR model with linear birth in the susceptible class and linear death at an equal rate in the removed class is analysed in Bailey [2, p. 136]. The assumption that all death in the fixed population N occurs in the removed class seems unreasonable. Moreover, for some parameter values the equilibrium point at which stability is analysed satisfies $I + S > N$ so that R would be negative, which is impossible.

REFERENCES

[1] Bailey, N. T. J. A simple stochastic epidemic, Biometrica 37, 193-202 (1950).

[2] Bailey, N. T. J. The Mathematical Theory of Epidemics, Charles Griffin, London, 1957.

[3] Brauer, F. and J. A. Nohel, _Qualitative Theory of Ordinary Differential Equations_, Benjamin, New York, 1969.

[4] Hethcote, H. W. Asymptotic behavior in a deterministic epidemic model, _Bull. Math. Biophys._, to appear.

[5] Hethcote, H. W. and P. Waltman, Optimal vaccination schedules in a deterministic epidemic model, _Math. Biosci._, to appear (1973).

[6] Weiss, G. H. and M. Dishon, On the asymptotic behavior of the stochastic and deterministic models of an epidemic, _Math. Biosci._ 11, 261-265 (1971).

RESILIENCE AND STABILITY AS SHOWN BY MODELS OF ECOLOGICAL SYSTEMS

C. S. Holling
Institute of Resource Ecology
University of British Columbia
Vancouver 8, B. C.

ABSTRACT

For the past ten years there has been an active application of systems methodology to ecological problems. From this has come a number of models of ecological processes, population interactions and ecological community structure. They range from models of a small number of state variables and a large number of parameters per variable, to those with thirty or more state variables and a small number of parameters per variable. Enough of these models have been developed to begin to show some common behavioural properties.

One of the more interesting concerns a general tendency of the modelled systems to exhibit more than one domain of "stability" or attraction around equilibrium points, trajectories or limit cycles, with at least one domain bounded by an unstable limit cycle. The feature of these boundaries is that they, rather than the area immediately surrounding the various equilibrium states, are critical to the overall behaviour of the system. Points on either side of the boundary will ultimately track to their respective predictable equilibrium states; points near the boundary are liable to be flipped across it from one domain to another in the face of small perturbations. The size of the domain, and the strength of the damping forces near its bounding edge, thus in large part characterize the ability of the system to maintain a structural integrity in the face of unexpected perturbations. If the domain is relatively small then a small perturbation can flip the system into another domain, thus altering its subsequent behaviour out of all proportion to the size and duration of the perturbation applied. Moreover, the weaker the damping forces in the vicinity of the boundary, the greater the likelihood that a small perturbation will cause that boundary to be crossed, regardless of the size of the respective domains. Finally, we note that in our ecological examples the parameter values occurring in nature seem generally to produce domains that are large, with rather weak damping around the equilibrium and strong damping at the boundaries.

From an equilibrium-oriented viewpoint, then, these systems can appear rather weakly damped and quite sensitive to disturbance. But from the viewpoint of the boundary, they are immensely stable with a high degree of persistence. In a sense this is what ecologists have always been saying - that what is important is not the efficiency of such systems, but the probability of their persistence. This orientation switches attention away from events near the equilibria to the events near the boundary of stability, and it is this switch that for us is placing so much of our understanding in a very new light.

We see some interesting consequences that could emerge by applying the resilience concept to policy analysis and the planning process. The analyses described above lead to the realisation that natural systems have experienced traumas and shocks over the period of their existence and the ones that have survived have explicitly been those that have been able to absorb these changes. They have, therefore, an internal resilience related to both the size of their domain of stability and the nature of the damping forces near the boundaries of the domain. So long as the resilience is great, unexpected consequences of an intervention of man can be absorbed without profound effects. But with each such intervention it seems that the price often paid is a contraction in the domain of stability until an additional incremental change can flip the system into another state. In a development scheme this would generate certain kinds of "unexpected" consequences in response to deceptively 'minor' perturbations - a freeway that changes the morphology of a city so that the urban core erodes; an insecticide that destroys an ecosystem structure and produces new pest species. We seem now to be faced with problems that have emerged simply because we have used up so much of the resilience of social and ecological systems. Up to now the resilience of these systems has allowed us to operate on the presumption of knowledge with the consequences of our ignorance being absorbed by the resilience. Now that the resilience has contracted, traditional approaches to planning might well generate unexpected consequences that are more frequent, more profound and more global. The resilience concept provides a way to develop a planning framework that explicitly recognises the area of our ignorance rather than the area of our knowledge.

The full paper entitled "Resilience and Stability of Ecological Systems" is published in Animal Review of Ecology and Systematics, Vol. 4, 1973.

THRESHOLDS FOR DETERMINISTIC EPIDEMICS

Frank Hoppensteadt
Courant Institute of Mathematical Sciences
New York University
New York, New York 10012

ABSTRACT

A general deterministic epidemic model is formulated which includes many of the deterministic models in the literature. The final size of epidemics described by this model (i.e., the number of susceptibles who eventually are exposed to the infection) is determined from an equation of the form

$$F = \exp [\gamma(F-1-\varepsilon)]$$

where ε is a measure of the size of the initial infectious population and γ is the number of initial susceptibles expected to be exposed to one infective. It is shown that the parameter γ has a threshold value one. If γ is less than one, few susceptibles will be exposed to the infection; while if γ is greater than one, a relatively severe epidemic will occur. This result reduces to similar threshold phenomena described by Kermack and McKendrick, Marchand, Wilson, Landau and Rapoport and the author for various special choices of the parameters of the general model.

Most models of epidemics reduce essentially to describing the interaction between two populations: the susceptibles and the infectives. The exposure rate of susceptibles is usually proportional to the number of infectives and the number of susceptibles. Those exposed to the infection may eventually become infectious themselves whereupon they may be quarantined, die or eventually recover from the infection. A quite general mathematical model can be formulated to describe the dynamics of such an epidemic.

The main interest in epidemic models stems from their use in uncovering certain qualitative features of epidemic processes. Of special interest is the final size of the epidemic (i.e., the number of susceptibles who become exposed to the infection) and its dependence on the various parameters characterizing the infection such as exposure rate, quarantine rate, length of the infectious period, etc.. One measure of the potential severity of an epidemic is the number of initial susceptibles who will be exposed to the infection by one infective. This is given by

$$\gamma = S_o r E$$

where S_o is the size of the initial susceptible population, r is the contact rate between infectives and susceptibles, and E is the life expectancy of infectives as infectives.

One expects that if there are few initial infectives and the number of susceptibles exposed by each infective is less than one, the infection will die out of the population with only a few susceptibles affected. On the other hand, if this number is greater than one, more new infectives will be created than there were initial infectives, and one expects that a severe epidemic will result. The point is that as γ passes through the value one, some change in the qualitative behavior of the final size should occur. Therefore, we refer to one as being the threshold value of γ.

For our model, E can be determined in the following way. If α is the rate at which infectives are removed through death and quarantine, etc., and σ is the length of the infectious period, then E is given by

$$E = \int_o^\sigma \exp\,[-\alpha x]dx,$$

and so

$$\gamma = S_o r \int_0^\sigma \exp\,[-\alpha x]dx = S_o r\,[1-e^{-\alpha\sigma}]/\alpha\ .$$

This expression for γ involves two dimensionless parameters, $S_o r/\alpha$ and $\alpha\sigma$, which are the essential parameters of the model.

The purpose of this note is to determine the final size of an epidemic and from it to demonstrate the threshold behavior for the general model. It will be shown that the final size can be determined by solving the equation

$$F = \exp\,[\gamma(F - 1 - \varepsilon)] \qquad (1)$$

where ε is a measure of the size of the initial infective population and $F = e^{\nu} S(\infty)/S_o$ where $\nu \geq 0$ is a constant to be defined later and $S(\infty)$ is the final size of the susceptible population. This equation can be solved numerically for various choices of the parameter γ, and the results are given in Figure 1.

FIGURE 1

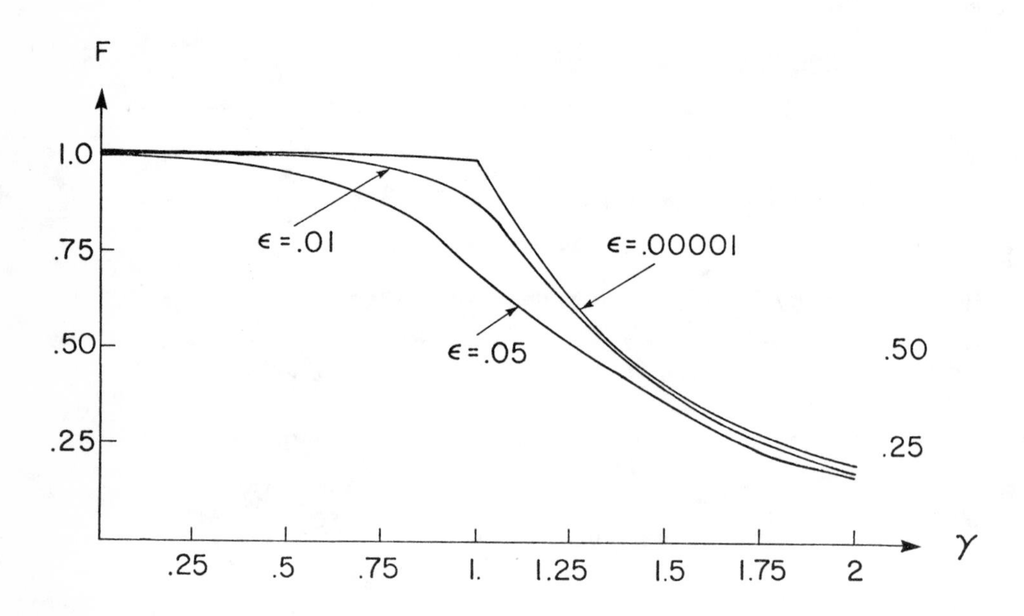

This calculation was carried out for the various choices of ε indicated on Figure 1 and for $\nu = 0$. The choices of ε near zero are the realistic ones since ε is roughly the ratio of initial infectives to initial susceptibles, and this typically is a small number. As expected, the importance of the threshold phenomenon decreases as the number of initial infectives becomes large.

For various choices of the model's parameters, equation (1) reduces to those discovered by Kermack and McKendrick [2], Marchand [4], Wilson [5], and in [3], to explain threshold phenomena in the particular models they considered, and it has been interpreted for age dependent models in the lecture notes [1].

We will now formulate the general model. The number of susceptibles at time t is denoted by $S(t)$, the infectives by $I(t)$. We first introduce a function $t - \tau(t)$ which is used to measure the holding time for individuals exposed but not yet infectious; i.e., an individual exposed at time $\tau(t)$ becomes infectious at time t. The function τ is a given non-decreasing function of time. Next, we assume no new susceptibles are introduced, and so the function

$$- dS(\tau(t))/dt$$

measures the rate at which new infections are arising at time t. And the function $\exp[\alpha(x-t)]$ gives the probability of a new infective surviving to age $t - x$. Therefore, the integral

$$- \int_{t-\sigma}^{t} [dS(\tau(x))/dx] \exp[\alpha(x - t)] H(x)dx$$

measures the number of new infectives in the population at time t. Here $H(x) = 0$ if $x < 0$, $= 1$ if $x > 0$, is used to ensure that only infections since time zero are accounted for. The total number of infectives in the population at time t then is

$$I(t) = I_o(t) - \int_{t-\sigma}^{t} [dS(\tau(x))/dx] \exp[\alpha(x-t)]H(x)dx \tag{2}$$

where the function $I_o(t)$ gives the number of original infectives remaining at time t. This is assumed to be known, and it is zero for $t > \sigma$.

As mentioned earlier, the exposure rate of susceptibles is taken to be proportional to I and S, so the equation governing the susceptible population is

$$dS/dt = -r\, I\, S\ , \qquad S(0) = S_o\ .$$

This equation can be integrated with the result

(3) $$S(t) = S_o \exp[-r \int_o^t I(x)\,dx].$$

The model we consider here is given by the two equations (2) and (3).

We note that $S(\infty)$ exists since $S(t)$ is a non-increasing function of t which is bounded below by zero. It will be shown that

(4) $$S(\infty) = S_o \exp\left[\gamma\left(\frac{S(\tau(\infty))}{S_o} - 1 - \mu\right)\right]$$

where

$$\mu = r\int_o^\sigma I_o(x)dx/\gamma .$$

If $\tau(\infty) = 0$, the final size is obviously $S_o \exp[-\gamma\mu]$ since no new infections can occur. Therefore, we consider the case

$$0 < \tau(\infty) \leq \infty, \text{ so } S(\tau(\infty)) = e^{\nu}S(\infty) \text{ for some } 0 \leq \nu \leq \gamma\mu .$$

If we set $F = e^{\nu}S(\infty)/S_o$, we have from (4) the equation

(5) $$F = \exp[\gamma(F - 1 - \mu + (\nu/\gamma))] .$$

Then (5) is just the equation (1) with $\varepsilon = \mu - (\nu/\gamma)$. It can be shown that $\nu \geq \gamma\mu$ cannot occur in the case $\tau(\infty) > 0$, and therefore we need only consider (1) with $\varepsilon > 0$.

It only remains to establish the equation (4). We write

$$\int_0^t I(x)dx = m(t) - n(t)$$

where

$$m(t) = \int_o^t I_o(x)dx,$$

and

$$n(t) = \int_{-\sigma}^{o} e^{\alpha y}\left\{\int_o^t [S(\tau(y+\hat{t}))]'\, H(y+\hat{t})d\hat{t}\right\} dy .$$

Since n can be rewritten for $t > \sigma$ as

$$n(t) = \int_o^\sigma e^{-\alpha u}\,[S(\tau(t-u)) - S_o]du,$$

we see that

$$m(\infty) = \int_0^\sigma I_0(x)dx$$

and

$$n(\infty) = \left(\int_0^\sigma e^{-\alpha y}dy\right)\ (S(\tau(\infty))-S_0)\ .$$

It follows that

$$S(\infty) = S_0 \exp\ [\ -\ r(m(\infty)\ -\ n(\infty))]\ .$$

But this expression equals

$$S(\infty) = S_0 \exp\ (\ r \int_0^\sigma e^{-\alpha x}dx\ \{S(\tau(\infty))-S_0-[m(\infty)/\int_0^\sigma e^{-\alpha x}dx]\})$$

which is equivalent to the desired result.

ACKNOWLEDGMENT

The author is grateful to D. Ludwig and J. B. Keller for helpful discussions and suggestions concerning these results. This research was supported by N.S.F. Grant No. GP-32996X2.

BIBLIOGRAPHY

1. Hoppensteadt, F. Lectures on population, epidemics and genetics.(in preparation)

2. Kermack, W.O., and McKendrick, A.G. A contribution to the mathematical theory of epidemics. Proc. Roy. Soc. A 115,700-721 (1927).

3. Landau, H.G., and Rapoport, A. Contribution to the mathematical theory of contagion and spread of information, I. Bull. Math. Biophysics 15, 173-183 (1953).

4. Marchand, H. Essai d'etude mathematique d'une forme d'epidemie. Ann. U. Lyons 3, 13-46 (1955).

5. Wilson, L. An epidemic model involving a threshold. Math. Biosci. (to appear)

DYNAMIC ANALYSIS IN "SOFT SCIENCE" STUDIES: IN DEFENSE OF DIFFERENCE EQUATIONS*

George Innis[†]
Natural Resource Ecology Laboratory
Colorado State University
Fort Collins, Colorado

ABSTRACT

The representation of dynamic systems with differential equations has a number of advantages. Among these are the well-developed theories on existence and uniqueness of solutions and on stability and convergence of approximate solutions. There are ***well-studied*** *numerical schemes implemented on many computers.*

In this paper three arguments are advanced for the use of difference equations for representing system dynamics in the "soft sciences." These are (i) clarity of communication with non-mathematicians, (ii) appropriateness of representation of soft science systems, and (iii) potential gain in knowledge and understanding from such representations. A simple Lotka-Volterra system is used to illustrate these points.

---

*This paper reports on work supported in part by National Science Foundation Grant GB-31862X2 to the Grassland Biome, U.S. International Biological Program, for "Analysis of Structure, Function, and Utilization of Grassland Ecosystems."

[†] Present address: Department of Wildlife Science, Utah State University, Logan, Utah.

1 INTRODUCTION

The principal concern of this paper is to advocate the use of difference rather than differential equations in the representation of the dynamics of "soft science" systems. For the purposes of this paper, soft science refers to biology, ecology, sociology, economics, political science, etc. These disciplines are distinguished by the difficulty involved in making precise measurements of the variables of interest. They are further distinguished because the laws which govern their dynamics are not known (in general) with great precision.

The calculus, of which differential equations are an offspring, is a tool of precision. It was conceived in the physical world, but born and raised in the mathematical world. This latter is characterized, among other things, by points (entities with location but without extent) and by point masses (points with a mass associated with them). This tool has certainly proven itself in the more precise applications. Engineering and physics have achieved heights unimaginable without the use of the calculus.

But the less precise the application, the less useful the tool; and, as I will show below, the more the tool may hinder rather than help.

This is the way of science however. Tools that are used to climb to one height can be burdens if carried beyond their range. We lose sight of this because the tool has been useful, because it is disloyal to abandon an old friend and helpmate, and because we become enamored by its bright sheen and forget that it is a tool. Nonetheless, we must remember what we are about. The tools which burden rather than support must be replaced.

2 COMMUNICATION

Many established and competent practitioners of the soft science disciplines are less than expert with calculus. They may have had no reason to study the material and even less reason to use it in their research. These people may have a great deal to offer to a dynamic analysis of the system. They know their system's normal dynamics through years of incisive study. They have seen many disturbances come and go and can describe the system's response to those disturbances. The descriptions often sound like "If the system looks like this (initial condition) and you do that (disturbance) to it, then after one week it will be here (state of the system) and after two weeks it will be here (state of the system)." As additional variables are considered, these scientists are able to describe the effects of these variables on the system dynamics.

This is a difference description of system dynamics. State descriptions are far apart (one week). The flow rates are long-term (one week) averages. The effects of additional variables are long-term considerations. Mathematically,

$$\underset{\sim}{x}(t+h) = \underset{\sim}{x}(t) + h\underset{\sim}{f}(\underset{\sim}{x}, \underset{\sim}{v}, t, h) \tag{1}$$

where $\underset{\sim}{x}$ is the state vector of the system, t is time, h is time step, $\underset{\sim}{f}$ is a rate vector describing the average rate of change of $\underset{\sim}{x}$ over the time interval $[t,t+h]$, and $\underset{\sim}{v}$ is a vector of other variables (exogenous) that affect the rates of change of $\underset{\sim}{x}$.

Thus, with this difference approach we can use the description of the dynamics directly in the equation structure. Likewise, the inverse transformation is easy, i.e., the difference equation can be read by the disciplinary scientist in his own words.

The dependence of $\underset{\sim}{f}$ on h is important. These scientists recognize that the description of the rate vector $\underset{\sim}{f}$ is different for different time steps h. Thus the description of rainfall in Kansas is different on an annual basis (seasonal distribution, average amount) from that used on an hourly basis (rainfall occurrence, amount). They are aware then that their description is not appropriate for arbitrary choices of h and, in particular, not appropriate in the limit as $h \to 0$.

To phrase equation (1) as a differential equation is mathematically simple if all the needed limits exist. Thus,

$$\frac{d\underset{\sim}{x}}{dt} = \underset{\sim}{f}(\underset{\sim}{x}, \underset{\sim}{v}, t, 0) \tag{2}$$

Unfortunately (or fortunately), this equation can be nonsense. Moreover, the time step h, important information for describing the process in (1), has disappeared in (2).

3 APPROPRIATENESS OF REPRESENTATION

One aspect of the appropriateness of the representation, i.e., the dependence of the $\underset{\sim}{f}$ in equation (1) on h, is sufficiently discussed in the preceding paragraphs. A second such point relates to the fact that $\underset{\sim}{f}$ is often (in the analysis of these systems) not well known. The rate function is chosen from a distribution of rate functions. At best we know the shape of the distribution. At worst, we have only vague notions about bounds on that distribution.

These considerations might suggest the use of stochastic differential equations. While such systems seem to have potential and are almost certainly more appropriate than the usual deterministic differential equations, there are other considerations. One of the most difficult points to treat is that system dynamics is often <u>grossly</u> independent of the rate function choice--provided that it comes from the given distribution. Stated another way, the state variables $\underset{\sim}{x}$ form such a self-regulating (homeostatic) system that variations in rate functions (or, experimentally, rate estimates) are compensated in other parts of the system. This compensation can take the form of damping or amplifying mechanisms elsewhere in the system and/or feedback from state variables to bring the given rate into line.

Describing such robustness mathematically is a challenge. Because $\underset{\sim}{f}$ is poorly known, the person doing the analysis has some latitude in choosing the way in which $\underset{\sim}{f}$ is represented. Some of these representations are brittle (e.g., positive feedback, recipient controlled systems); and others are robust (e.g., negative feedback, donor controlled systems). The analyst often has little reason from the viewpoint of representing a given rate function's dependence on a vector of controlling variables [$\underset{\sim}{x}$ and $\underset{\sim}{v}$ in (1)] to choose a given rate function from the class. If he considers the homeostasis of the system, however, then he may well find that some functional forms available are far more "robust" than others. The example presented below contains illustrations of a robust and a brittle representation of the same system that are equally defensible from the viewpoint of describing the rate function.

A third consideration for the appropriateness of difference vs. differential equations harks back to the limit concept. The Heisenberg Uncertainty Principle (Richtmyer and Kennard 1947) assures us that we cannot simultaneously resolve both time and space with arbitrarily great accuracy. Thus,

$$h \, \Delta z \geq \hbar$$

where Δz is the separation of points that are resolved in time h and $\hbar$ is Planck's constant ($\sim 6.6 \cdot 10^{-27}$ μg-sec). This point assures us that differentiation, integration, and other limit processes are mathematical concepts without literal equivalents in the physical world; i.e., there do not exist physical processes for determining derivatives of physical variables. This does not mean that physical quantities do not vary, but that the instantaneous rate of variation (derivative) of the variable is not measurable. The average rate of variation over some time interval is, of course, physically determinable (at least approximately).

The smallness of Planck's constant might cause us to wonder if the Heisenberg Uncertainty Principle is important in the practical arena of interest. A further consideration of equation (2) will lead to more insight on this problem. In (2) the derivative of $\underset{\sim}{x}$ depends upon $\underset{\sim}{x}$. For this to occur physically, instantaneous values of $\underset{\sim}{x}$ must be sensible to the physical system. Systems of the type discussed here are typically spatially distributed with $\underset{\sim}{x}$ representing only the average value of the state variables over some spatial regime. Clearly, time is necessary to determine this average. Time is required for the system to sense a change in the average and thus the derivative of $\underset{\sim}{x}(t)$ does not depend on $\underset{\sim}{x}(t)$.

Contrast this with equation (1) which indicates that the state in the future, $\underset{\sim}{x}(t+h)$, depends on the state now, $\underset{\sim}{x}(t)$, and on some average rate for the interval $[t,t+h]$. If h is larger than the time that the system takes to sense its state $\underset{\sim}{x}$, then with $\underset{\sim}{f}$ chosen appropriately, (1) seems a reasonable description of the way the system changes state. See the section on difference equation solutions for implications on existence.

The preceding paragraphs cast doubt on the representation of any density dependent mechanisms in spatially distributed systems using ordinary differential equations. A simple demonstration of this fact is easily constructed from the Lotka-Volterra system described below. Consider first such a system operating in a laboratory container. One dynamic system attains. Next consider the container as subdivided by a membrane with different initial densities on either side of it. If the membrane is less than perfectly permeable, the dynamics of the system is different from that of the undivided container. Even if one accepts the Lotka-Volterra representation of system dynamics, this result is not at all surprising. The point is that in spatially distributed systems, it is the local density to which the system responds (if it responds to a density at all). Thus consistent application of differential equation concepts would imply partial differential equations even in the simplest cases of non-uniform distribution of variables.

4 POTENTIAL GAIN IN KNOWLEDGE AND UNDERSTANDING

Suppose the dynamics predicted by the differential equation system in (1), as determined by solution on a digital computer, are inconsistent with known or expected system dynamics. What are the potential sources of error?

In Fig. 1 we illustrate the steps involved in the solution process. Following the problem statement the system is usually described in difference terms using the available laws (of conservation, for example). This step may be omitted in problems where the differential representations are known. For digital numerical treatment, the differential equations are reduced to difference equations, and an appropriate numerical solution scheme is applied. The results of the numerical solution are compared with a precision indicator; and if the desired precision is not obtained, a refinement of the difference representation is tried. This refinement proceeds until the desired precision is achieved.

There is the potential for error, of course, at each step of this torturous process. The difference equations may fail to contain all of the important effects for achieving the goals of the present analysis. The limit process for deriving the differential equations may destroy information. The difference approximation to the differential equations may be inappropriate. Numerical solutions accumulate their own sorts of errors. The analyst is faced with a multitude of options for explicating any discrepancy between the numerical solution and the known or expected system dynamics.

The science of numerical analysis appropriately takes the differential equation system in (2) as given, and seeks efficient, stable numerical techniques for finding

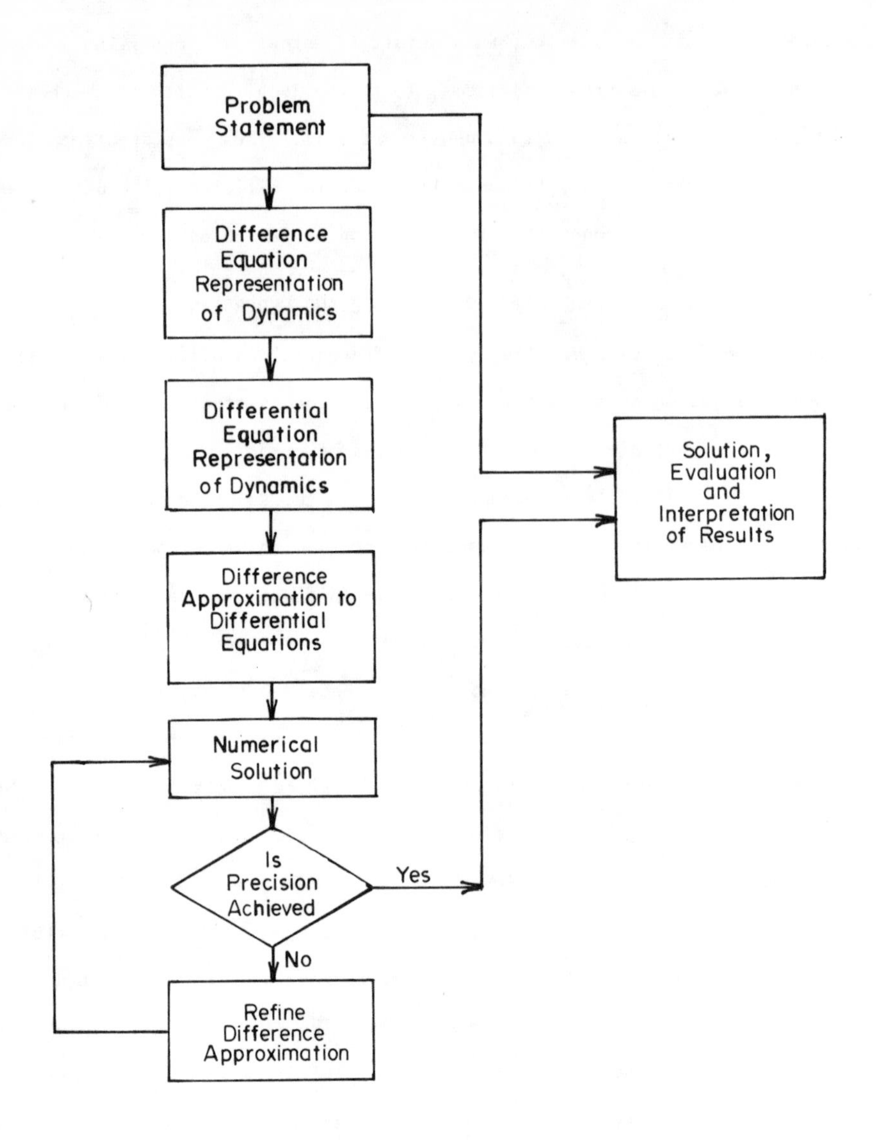

Fig. 1. An overview of steps involved in the dynamic analysis of a problem using differential equations which are solved on a digital computer.

approximate solutions to these equations. If, as argued above, the differential equation representation of the system dynamics are imprecise to begin with, the attention given the search for precise solutions may be inappropriate if not misleading. In the Lotka-Volterra example below, the existence of an exact (closed form) solution to a set of approximate equations has led, it is argued, to unwarranted attention being given to this system.

Contrast this with Fig. 2 wherein the solution of a problem using difference equations is depicted. In this case the analyst knows that the difference equations that are numerically solved (modulo coding errors) are the ones used to describe the system dynamics on the time scale of interest. If there are discrepancies apparent in the solution evaluation, these can only stem from an inappropriate representation of the system dynamics. Thus the analyst is provided with stronger and more direct feedback to the representation of the dynamics than he is in the differential equation case. This point is illustrated in the example below by the fact that the Lotka-Volterra system is unstable in difference form. The analyst recognizes this to mean that the given representation of the biological processes is incorrect. One additional hypothesis is described there which stabilizes the system.

5 THE LOTKA-VOLTERRA EXAMPLE

To illustrate the generalities above, consider the Lotka-Volterra system of Fig. 3 and equations (3).

$$\frac{dH}{dt} = (A_1 - B_1 P)H$$
$$\frac{dP}{dt} = (A_2 H - B_2)P \qquad (3)$$

This figure is a Forrester (1961) diagram illustrating the interactions operative in the simple Lotka-Volterra system. This is an extensively studied system. For initial studies of Lotka and Volterra, see Lotka (1925). Peilou's (1969) treatment is particularly thorough and illustrates a class of studies that have been conducted on this system. A difference equation representation of the same system is given in equations (4):

$$H(t+h) = H(t) + h[A_1 - B_1 P(t)]H(t)$$
$$P(t+h) = P(t) + h[A_2 H(t) - B_2]P(t) \qquad (4)$$

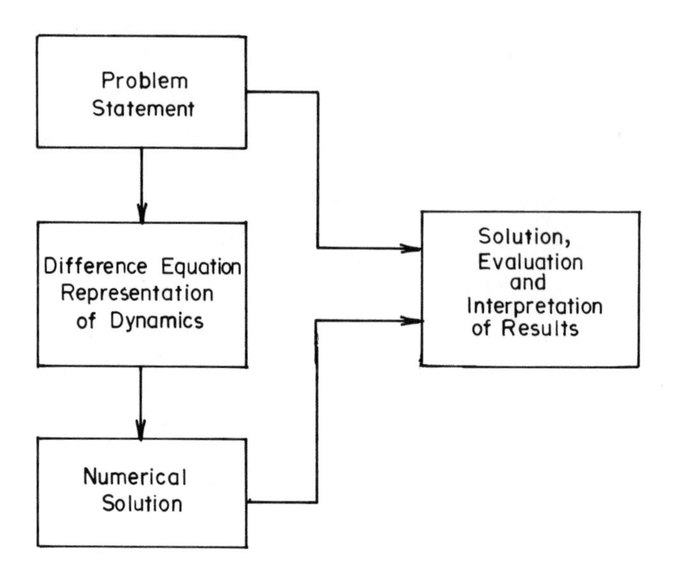

Fig. 2. An overview of steps involved in the dynamic analysis of a problem using difference equations which are numerically solved on a digital computer.

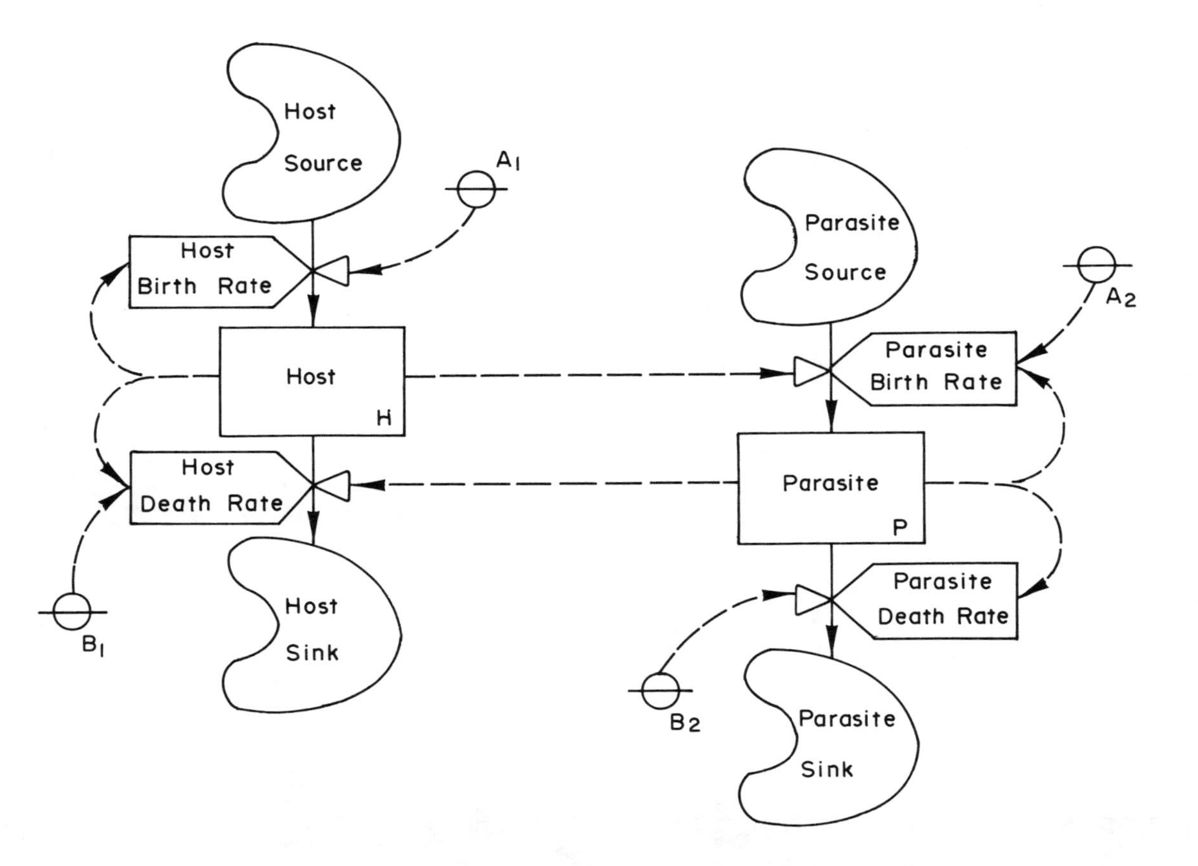

Fig. 3. A Forrester diagram of a simple Lotka-Voltera System.

While it is clear that mathematically (4) is just an Euler approximation (Hildebrand 1956) to (3), it is also clear that (4) may be conceptually different from (3).

If (4) is derived by the biologist describing the average dynamics over the time interval h for the populations H and P, then (4) is appropriate and (3), derived by doing a little algebra and taking limits in (4), is inappropriate. If the difference between these two systems is not carefully explained or understood, then the fact that the solution to (3) is presented as a solution to (4) can be most misleading.

In Fig. 4 the solution to (3) is presented for a given choice of the parameters and initial conditions. The biologist who, for the sake of discussion, formulated (4) was seeking a description of the dynamics of these populations which displayed cyclic behavior since, grossly, this is observed naturally. He is not aware of all the mechanisms operating between the real-world populations. The modeling study is intended to investigate the properties of the hypotheses needed to display such dynamics. Having been shown Fig. 4, he concludes that (4), which the mathematician has played with a little bit in converting it to (3), is a description of the interactions of these populations which displays the desired dynamics. He proceeds to a series of further analyses based on (4) as an appropriate conceptualization of the interrelationships between H and P.

The biologist has been misled. The solution to (4) is shown in Fig. 5 for the same initial conditions and parameter values used in Fig. 4. However, instead of an h of zero (a differential equation solution), we have in Fig. 5 an h of 0.2 (which for the sake of discussion we will take as that h which was part of the biologist's conceptualization of the problem). Many mathematicians would not be surprised by the shape of this curve, but would recognize immediately that the time step in the Euler solution to (3) was too large. If they choose to help the biologist by correcting this error they will have destroyed information that the biologist could have used to improve his conceptualization of the host/parasite system.

If the biologist analyzes Fig. 5 as the relationship between host and parasite which is obtained from his conceptualization of that system [equation (4)], he

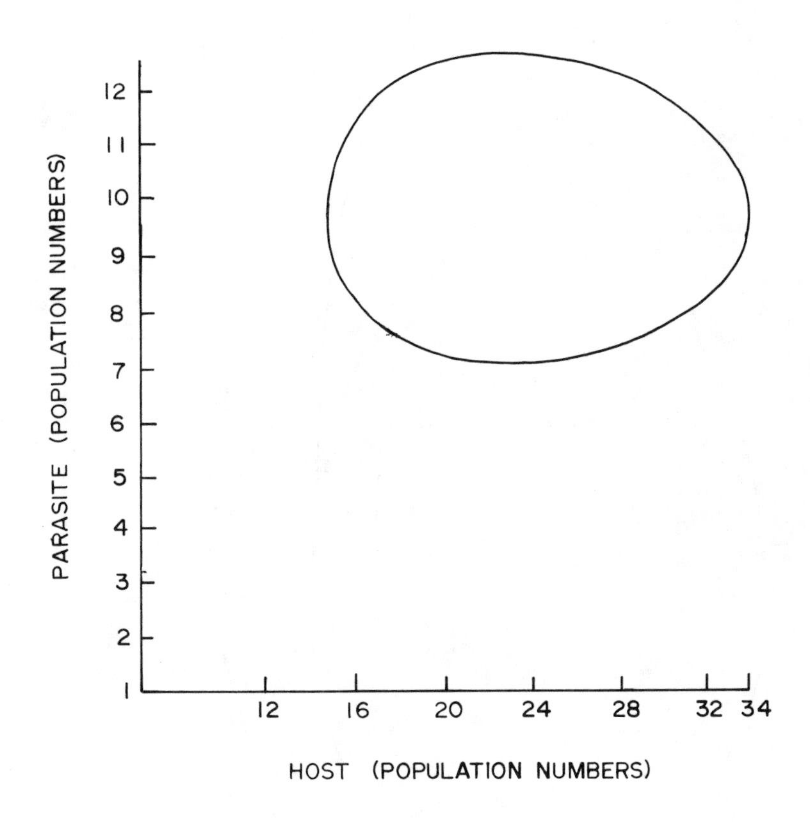

Fig. 4. Solution of equation (3), the Lotka-Volterra system, with $A_1 = 1.$, $B_1 = 0.1$, $A_2 = 0.02$, $B_2 = 0.5$, $H(0) = 20.$, $P(0) = 7$.

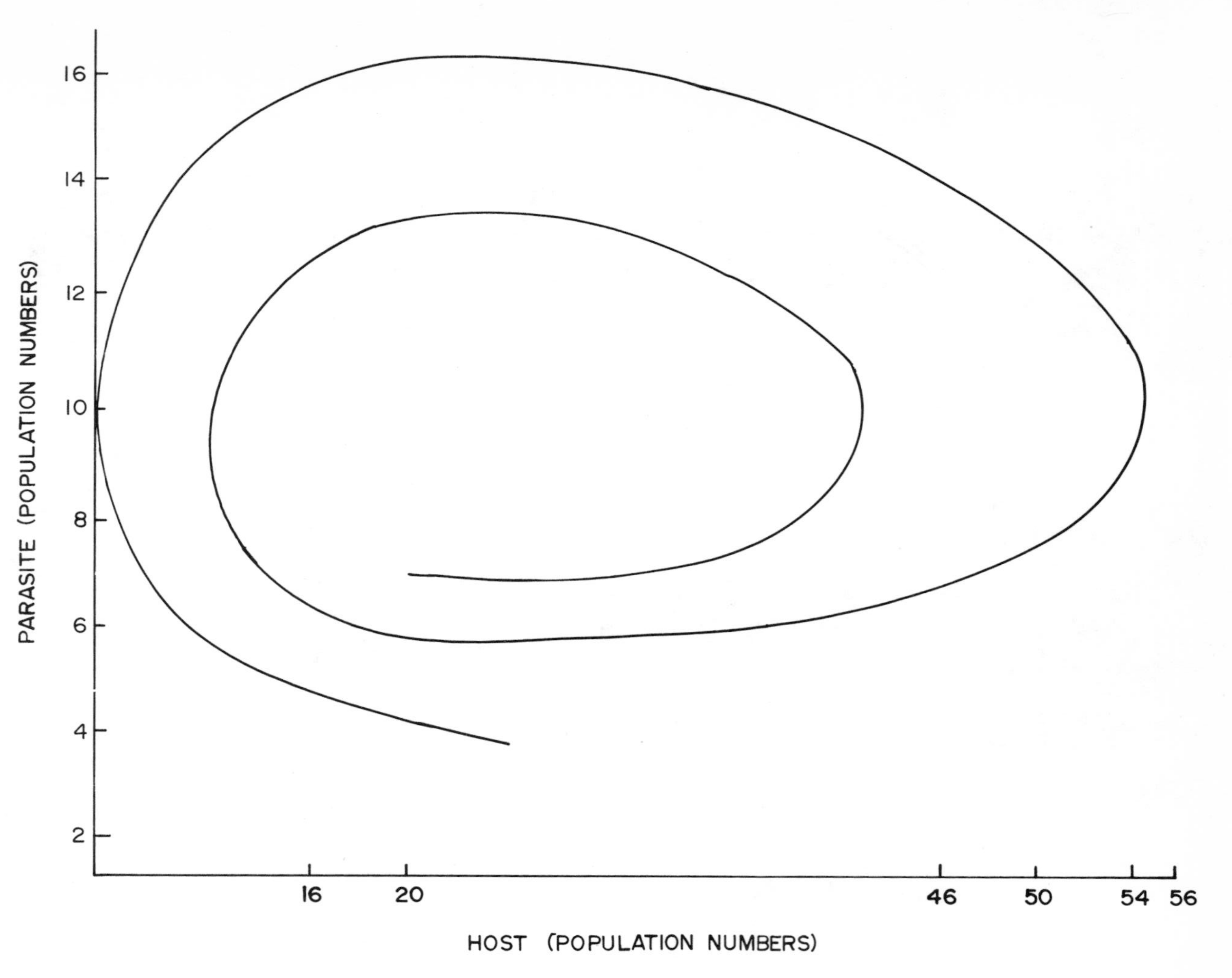

Fig. 5. The solution of equation (4) with $A_1 = 1.$, $B_1 = 0.1$, $A_2 = 0.02$, $B_2 = 0.5$, $H(0) = 20.$, $P(0) = 7.$, and $DT = 0.2$.

realizes that other biological mechanisms must be operational within that system to achieve cyclic (or approximately so) dynamics. He may know a great deal more about the interactions between host and parasite. He may recognize that, for example, host relative reproduction rate (numbers produced per individual in the population per unit time) rises as the population falls, and perhaps that parasite relative death rate increases as that population increases. The increased birth rate may be explained by the greater availability of reproduction sites (nests, dens, burrows, etc.), and the increased death rate may result from the stress of behavioral interactions when the populations become too dense. Incorporating these concepts into the problem, Fig. 6 is the revised Forrester diagram, equations (5) are the difference equations, and Fig. 7 illustrates the dynamics that are obtained.

$$\begin{aligned} H(t+h) &= H(t) + h[A_1 + C_1(HPN - HLT) - B_1P(t)]H(t) \\ P(t+h) &= P(t) + h[A_2H(t) - B_2 + D_2(PPN - PLT)]P(t) \end{aligned} \tag{5}$$

As Fig. 7 illustrates, this system provides approximately cyclic dynamics as the biologist expects and incorporates a bit more of the biologist's knowledge of host and parasite interactions.

There are, of course, many alternative hypotheses that would result in similar cyclic (or approximately so) dynamics. One of the most interesting is derived from taking a very stable (in the sense of rapidly converging to the equilibrium value of $H = 25.$, $P = 10.$) choice of parameters in equations 5 such as $A_1 = 1.$, $B_1 = 0.1$, $A_2 = 0.02$, $B_2 = 0.5$, $C_1 = 0.01$, $D_2 = 0.005$. The relatively large values of C_1 and D_2 cause the spiral of Fig. 5 to reverse and tend toward equilibrium. However, after computing the coefficients of h in equations 5, these coefficients are multiplied by pseudorandom numbers uniformly distributed in [0.5, 1.5]. The resulting host/parasite dynamics are roughly cyclic, but the random perturbations provide a biological realism not found in the deterministic representations. Further efforts with this sort of stochastic difference equation have yielded fascinating results that will be reported elsewhere.

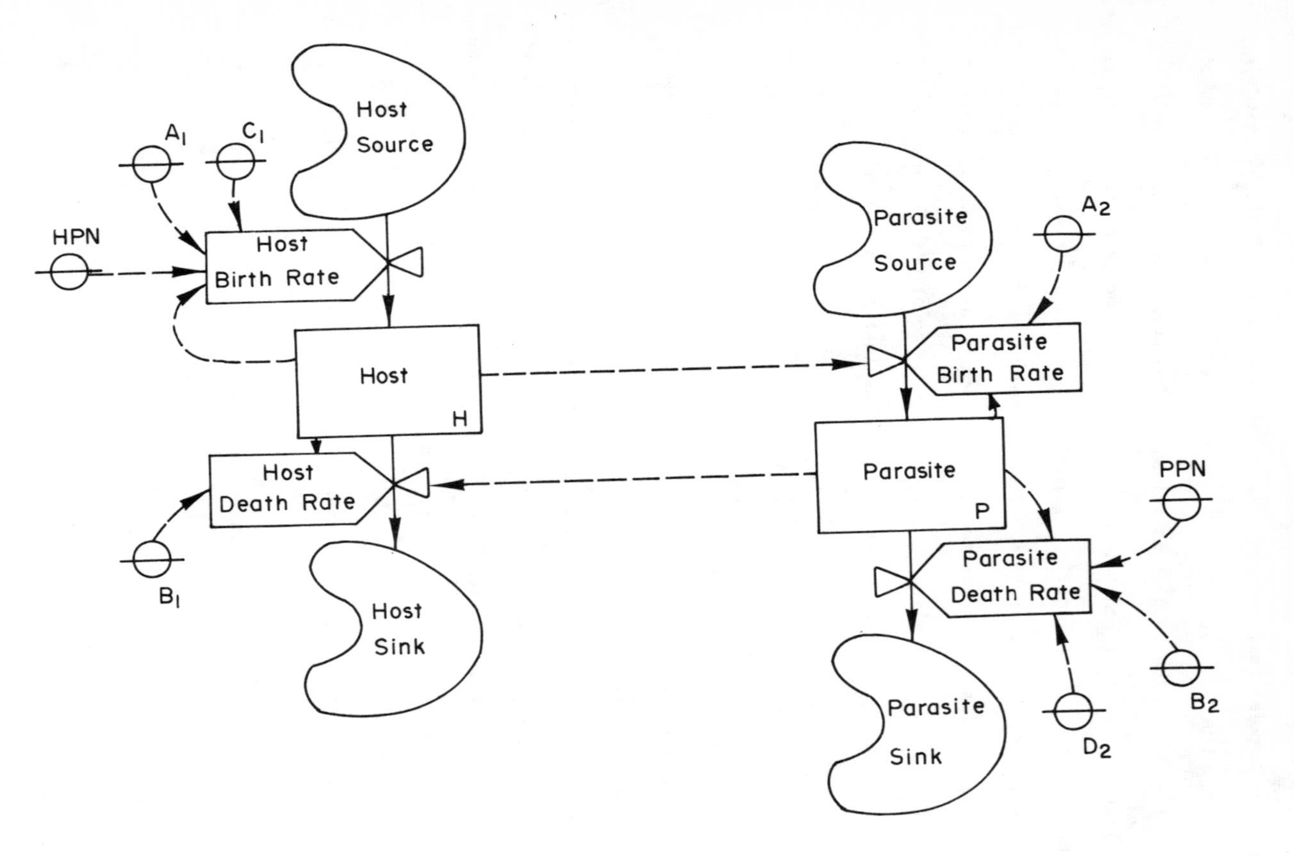

Fig. 6. One modification of Fig. 3 that might be obtained if the biologist had seen the results in Fig. 5.

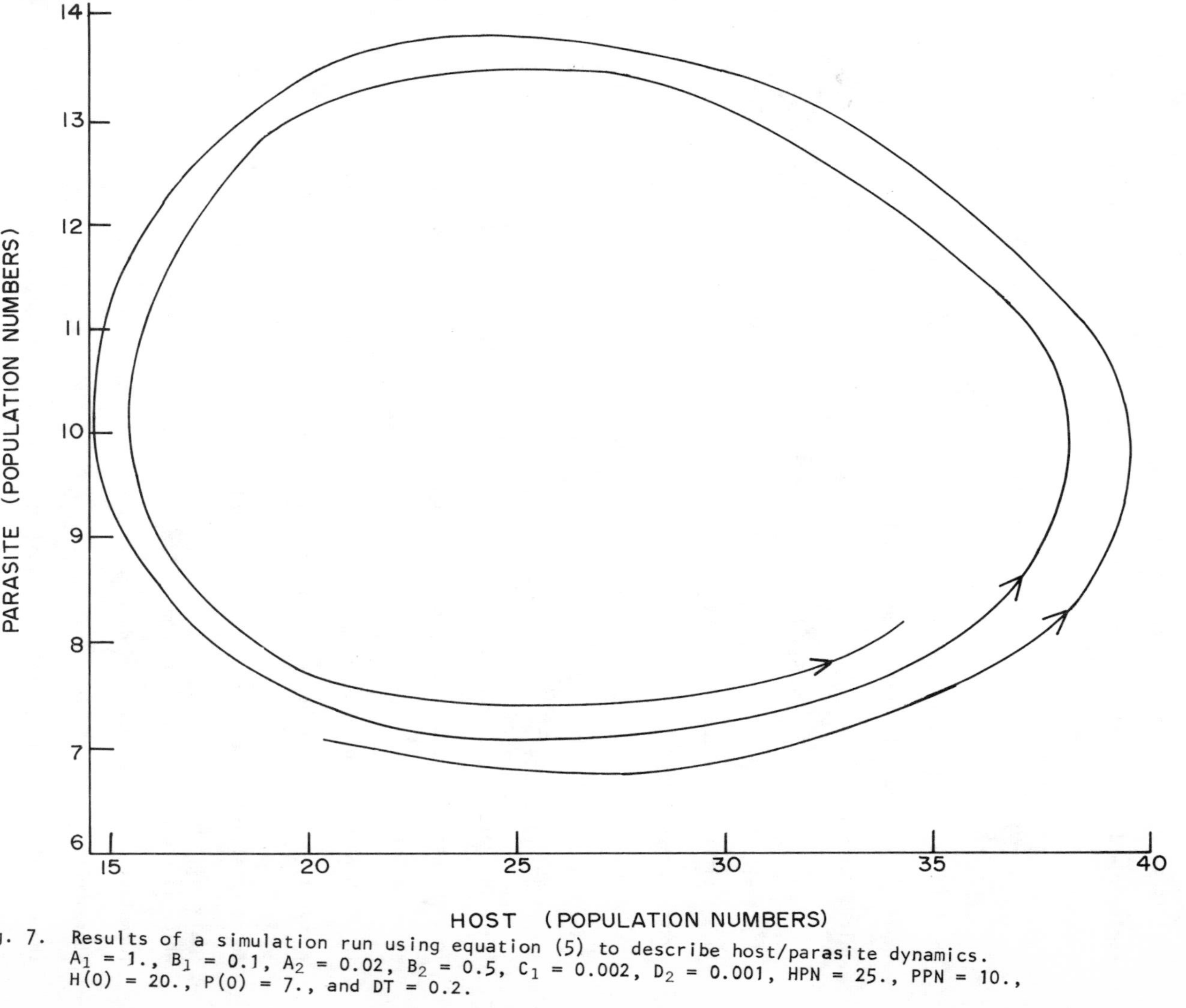

Fig. 7. Results of a simulation run using equation (5) to describe host/parasite dynamics. A_1 = 1., B_1 = 0.1, A_2 = 0.02, B_2 = 0.5, C_1 = 0.002, D_2 = 0.001, HPN = 25., PPN = 10., H(0) = 20., P(0) = 7., and DT = 0.2.

6 DIFFERENCE EQUATION SOLUTIONS

The general problem of existence and uniqueness for the solution of first-order difference equations is unsolved (Bellman 1947). Problems of interest in biology, ecology, and industrial dynamics, and other such disciplines can all be written in the form

$$\Delta z_i = z_i(t+h) - z_i(t) = hf_i[z_1(t), z_2(t), \ldots, z_N(t), t]$$

$$i = 1, 2, \ldots, N; \; t = 0, h, 2h, \ldots$$

where the z_i are the state variables of the system. This particular system of non-linear first-order difference equations has a unique solution corresponding to each initial condition.

$$z_i(0) = C_i \qquad i = 1, 2, \ldots, N$$

if the z_i are unique (i.e., z_i and z_j cannot be the same variable satisfying different difference equations). Existence and uniqueness of the solutions of such systems are clear (Bellman 1947).

A good bit of discussion has centered on the dependence (or lack of it) of one rate on another in biological and physical systems. This author (Innis 1972) and others (Forrester 1968*a*) have written that such dependence is physically impossible. It is interesting that the absence of such dependence is important for existence. To elaborate, note that the dependence of one rate on another can be a trivial expedient or a fundamental consideration. A trivial expedient refers to the situation in which each rate depends in a similar way on state and driving variables and, as a result, one rate can be written as a function of the other. If the dependence is fundamental, then the rate over the interval t to t+h is expressed as $\dfrac{x_i(t+h) - x_i(t)}{h}$ which is the average rate of change of x_i over the interval from t to t+h. In this formulation f_j, the rate dependent on this rate, can be expressed as a function of states only, but now f_j is a function of states at time t <u>and at time t+h</u>. Under these conditions, more restrictions must be placed on f_j to guarantee existence as the following examples indicate. Suppose

$$x(t+h) = x(t) + hf[x(t), x(t+h)]$$

$$x(0) = 0$$

$$f(x,y) = \begin{cases} 1 \text{ if } x = y \\ 0 \text{ if } x \neq y. \end{cases}$$

Now

$$x(h) = x(0) + hf[x(0),\, x(h)]$$

$$= hf[0,\, x(h)].$$

$$\text{If } x(h) \neq 0, \text{ then } f[0,\, x(h)] = 0 \Longrightarrow x(h) = 0.$$

$$\text{If } x(h) = 0, \text{ then } f[0,\, x(h)] = 1 \Longrightarrow x(h) = h \neq 0.$$

These contradictions imply that no solution exists.

As a second example, consider

$$f(x,y) = \frac{y}{h}$$

$$x(0) = a \neq 0.$$

Then

$$x(h) = x(0) + h \cdot \frac{x(h)}{h}$$

or

$$x(h) = x(0) + x(h).$$

For $x(0) \neq 0$, there is no choice for $x(h)$ which satisfies the last equation.

7 CHARACTERISTICS OF THE SOLUTIONS

Consider the set $\{0, h, 2h, \ldots\} = I_h$ as a subset of the real line with the discrete topology. If each of the f_i is continuous in each argument, then each x_i is continuous on I_h. If, as is the case in ecological studies, the interval of interest is finite, then if $L_h = [0,L] \cap I_h$, each x_i is uniformly continuous on L_h and therefore each x_i is unconditionally stable on L_h.

This is a result of interest in biological modeling because it assures us that small variations in initial conditions results in small variations in the dynamics of the state variables.

CONCLUSION

The mathematically inclined reader may ask if we have not delved deeply into the intrinsically obvious. The disciplinary scientist (non-mathematician) may wonder if the mathematician is not playing games again. Let me conclude by drawing some implications from these arguments and by asking a few questions.

1. Do modelers in the soft sciences need to know the calculus (limit)? The arguments above indicate that the calculus and differential equations are unnecessary

for the construction of dynamic models of such systems. If the model is conceived, written, and analyzed in difference terms, the limit concept never arises.

2. Why study differential equations? There are several answers. (i) One reason is historical; modeling in these soft science areas is new and is taking its lead from models built in engineering and physics. (ii) The use of differential equations allowed for the solution of many interesting problems that were completely unmanageable prior to digital computers. (iii) We do not know as much about difference equations (closed form solutions, stability characteristics, etc.) as we do about differential equations. This point is important in choosing functional representations, checking complex computer programs that purport to solve difference equations, and developing difference equation tools patterned after the differential equation counterparts. (iv) Many useful models are written in differential equations.

3. Forrester (1968*b*) claims that without the experience of the analysis of engineering feedback loop system (or some equivalent) one can build such models (difference equation), but one cannot appreciate their nuances. This is probably more commentary on current approaches to teaching modeling than an indication of an intrinsic feature of model building and analysis.

4. As long as a genuine understanding and facility with calculus is required to build dynamic models, the club of model builders has some real job security.

5. Many modelers in the soft science areas do not possess sufficient command of their quantitative skills to vary the applications much beyond their educational examples. This problem existed in engineering for a long period (and, indeed, exists still, though to a lesser extent).

6. The experience of almost all of us in science is to make precision a goal to be striven for in all things scientific. Einstein is said to have described his mental pictures while developing the theory of general relativity as "clouds." Fuzzy sets (Zadeh 1965), statistical mechanics, and other such studies may serve as a basis for dealing with the imprecision of some disciplines. The mathematical concepts of calculus, differential equations, limits, etc., have as an implicit component a precision that transcends any physical object or science.

7. Finally, the advancement of any science is a cyclic process. Each turn round the cycle moves us higher toward knowledge and understanding. At each turn we must, among other things, be sure that our tools are furthering our objectives, not burdening us with dead weight. This is hard to do, and it is risky. However, failure to put the obsolete aside may be even more troublesome.

REFERENCES

Bellman, R. 1947. On the boundedness of solutions of nonlinear differential and difference equations. Amer. Math. Soc., Trans. 62:357-386.

Forrester, J. W. 1961. Industrial dynamics. MIT Press, Cambridge, Massachusetts. 464 p.

Forrester, J. W. 1968*a*. Principles of systems. Wright-Allen Press, Cambridge, Massachusetts. 300 p.

Forrester, J. W. 1968*b*. Industrial dynamic - after the first decade. Manage. Sci. 14(7):398-415.

Hildebrand, F. B. 1956. Introduction to numerical analysis. McGraw-Hill Book Co., Inc., New York. 511 p.

Innis, G. S. 1972. The second derivative and population modeling: Another view. Ecology 53(4):720-723.

Lotka, A. J. 1925. Elements of physical biology. Williams and Wilkins, Baltimore, Maryland. 460 p.

Pielou, E. C. 1969. An introduction to mathematical ecology. Wiley-Interscience, New York. 286 p.

Richtmyer, F. K., and E. H. Kennard. 1947. Introduction to modern physics. 4th ed. McGraw-Hill Book Co., Inc., New York. p. 267.

Zadeh, L. 1965. Fuzzy sets and systems, p. 29-37. *In* Symposium on system theory. Polytechnic Press, Polytechnic Inst., Brooklyn, New York.

Probabilistic Limit Cycles

Clark Jeffries

Introduction

Populations change. Some natural populations seem to change in a regular, cyclic manner. This report is about models of populations which so change or fluctuate. We develop a rather simple mathematical idea: Suppose a Lotka-Volterra model of a predator-prey system enjoys asymptotic stability for some population combination. However, suppose that this stability is weak in the sense that the populations, when displaced slightly from equilibrium, return slowly to levels near equilibrium and return only after oscillating around equilibrium "many" times. If such a system is given small, frequent, random perturbations it will tend not to return to equilibrium at all, but will oscillate around equilibrium in a fairly regular way, indefinitely.

Predator-Prey Systems

The interaction between a prey population x and a predator population y is, of course, a classical topic in the mathematical ecological literature. R. M. May (1972) has reviewed a class of predator-prey dynamical systems of the form

$$dx/dt = x\, g(x,y) \tag{1a}$$

$$dy/dt = y\, h(x,y) \tag{1b}$$

and has presented criteria for the functions g and h so that the system will exhibit asymptotic stability at a point or a stable limit cycle (see Figure 1 of May (1972)). For more details on such models see Albrecht, Gatzke, and Wax (1973), Gilpin (1972), and Scudo (1971).

We consider a simple extension of the classical Lotka-Volterra model:

$$dx/dt = x\,(-Ax - By + E) \tag{2a}$$

$$dy/dt = y\,(Cx - Dy - F) \tag{2b}$$

where A, B, ... , F are all positive constants. The ascription of ecological signigicance to the coefficients should be familiar to most readers.

A critical point for Equations 2a, 2b (where $dx/dt = 0$ and $dy/dt = 0$) may be found in positive population space (all combinations (x, y) such that $x > 0$ and $y > 0$) if and only if $E/A > F/C$. This is a sort of minimal production-consumption condition. If the system meets this condition, then the populations may be normalised (or "weighted") so that (1, 1) is a critical point. Henceforth it is assumed that the model has been so normalised, so $A + B = E$ and $C - D = F$.

Thus we may rewrite Equations 2a, 2b as

$$dx/dt = x\,[-A(x-1) - B(y-1)] \quad (3a)$$

$$dy/dt = y\,[C(x-1) - D(y-1)] \quad (3b)$$

or in a mathematically simpler form

$$dx/dt = -A(x-1) - B(y-1) \quad -A(x-1)^2 - B(x-1)(y-1) \quad (4a)$$

$$dy/dt = \underbrace{C(x-1) - D(y-1)}_{\text{linear terms}} \quad \underbrace{+C(x-1)(y-1) - D(y-1)^2}_{\text{second order terms}}. \quad (4b)$$

Sufficiently close to (1, 1) the linear terms of Equations 4a, 4b are dominant. The second order terms are important elsewhere. The Linearisation Theorem of differential equations theory may be applied, and it follows that the above normalised model is asymptotically stable at (1, 1) (see R. Rosen (1970) pp. 117-120 and Rosenzweig and MacArthur (1963)). Technically, asymptotic stability at (1, 1) is guaranteed by the Linearisation Theorem because the real parts of the eigenvalues of the linear approximation matrix

$$\begin{pmatrix} -A & -B \\ C & -D \end{pmatrix}$$

are negative.

It may be shown that, using Equations 3a, 3b, any initial positive population vector will follow a trajectory which will asymptotically approach a limit, namely (1, 1) . So we say that the <u>basin of attraction</u> of (1, 1) is all of positive population space.

For the remainder of this section we will illustrate probabilistic limit cycles by refering to variations of the following explicit model:

$$dx/dt = x\,(-.1x - 2y + 2.1) \quad (5a)$$

$$dy/dt = y\,(2x - .1y - 1.9). \quad (5b)$$

The mathematical significance of the coefficients here is that (1, 1) is a critical point and that the real parts of the eigenvalues of the linear approximation matrix are negative. In fact, the real parts of the eigenvalues are both $-.1$ and hence small in magnitude when compared with the imaginary parts, namely $\pm 2\sqrt{-1}$. Therefore the dynamical system given by Equations 5a, 5b is only weakly damped; a population trajectory starting from, say, (2, 2) would spiral "many times" while approaching (1, 1).

A computing machine may be programmed to approximate, using Euler's method, the trajectories associated with Equations (5a), (5b); and a convenient time interval for iterations is .01 time units. If at each iteration two uniformly random numbers distributed between $\pm.07$ are multiplied by and then added to the populations, a sort of limit cycle behavior results. That is, a <u>probabilistic limit cycle</u> is associated with the difference equations

$$\Delta x = x(-.1x - 2y + 2.1)\Delta t + pert_1 \cdot x \qquad (6a)$$

$$\Delta y = y(2x - .1y - 1.9)\Delta t + pert_2 \cdot y \qquad (6b)$$

where $\Delta t = .01$ and where $pert_1$ and $pert_2$ are uniformly random with range $\pm .07$. See Figures 1, 2, and 3.

We may characterise this probabilistic limit cycle (somewhat arbitrarily) by noting where and when the trajectory crosses the $x = y$, $x, y < 1$ line. The average value at crossing is about .5 population units and the average period between crossings is about 3.6 time units. See Figures 4 and 5.

The obvious biological significance of Equations 5a, 5b is that population changes are much more dependent upon interspecific interactions than upon self-crowding at populations near (1, 1). Probabilistic aspects further characterise the model's biological implications. As depicted in Figures 4 and 5, there is considerable variation in the frequency and amplitude of the cycles. Particularly long or short cycles are associated with particularly large or small amplitudes, respectively. Infrequently the population vector may approach (1, 1) where the trajectory becomes temporarily almost random. Thereafter the trajectory drifts by chance away from (1, 1) and resumes oscillating. Cycles with small amplitudes tend to occur consecutively, as do cycles with large amplitudes.

The global stability of Equations 6a, 6b is also probabilistic. If the system is displaced away from (1, 1) , say, to (.01, .01) or (20, 20) , then the system will recover in the sense that it will fall into the typical probabilistic limit cycle pattern within a few time units. However, displacing the system very far from (1, 1) , say, to (100, 100) will lead to collapse of the system. Establishing lower population thresholds would lead to a model incapable of withstanding wholesale enrichment or decimation.

If we "eliminate" predators and consider the difference equation

$$\Delta x = x(-.1x + 2.1)\Delta t + pert_1 \cdot x \qquad (7)$$

then we find that this isolated prey system is again prone to fluctuations around a new equilibrium level, $x = 21$. The prey population will fluctuate readily from <15 to >30 population units. If a natural predator-prey system could be approximated by Equations 6a, 6b and if predators were eliminated, then observers would expect the prey population to continue to fluctuate, only at higher population levels. We have observed some computer simulations of this which could be mistaken

(continued page 6)

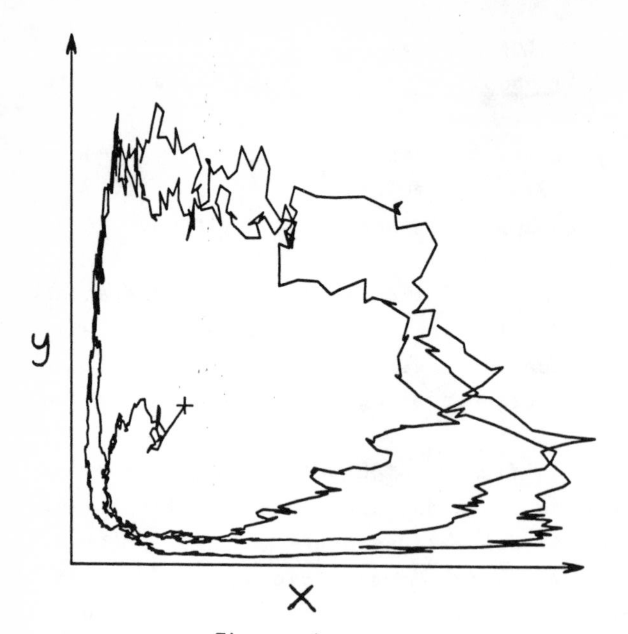

Figure 1
This is a computer generated trajectory associated with Equations 6a,6b. The trajectory begins at (1, 1) , here denoted by "+" . The time span is 10 time units or 1000 iterations. About two and one half well defined cycles are indicated.

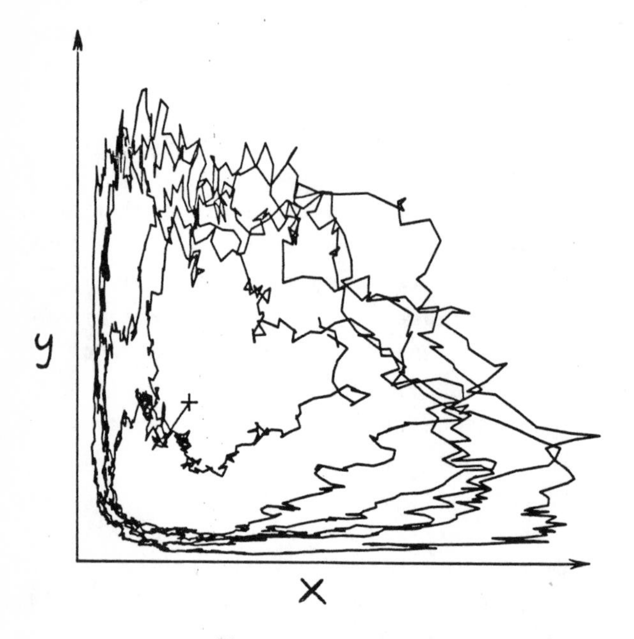

Figure 2
A continuation of the graph in Figure 1 is shown here up to 20 time units. Three cycles have been added, and there is considerable variation in amplitude.

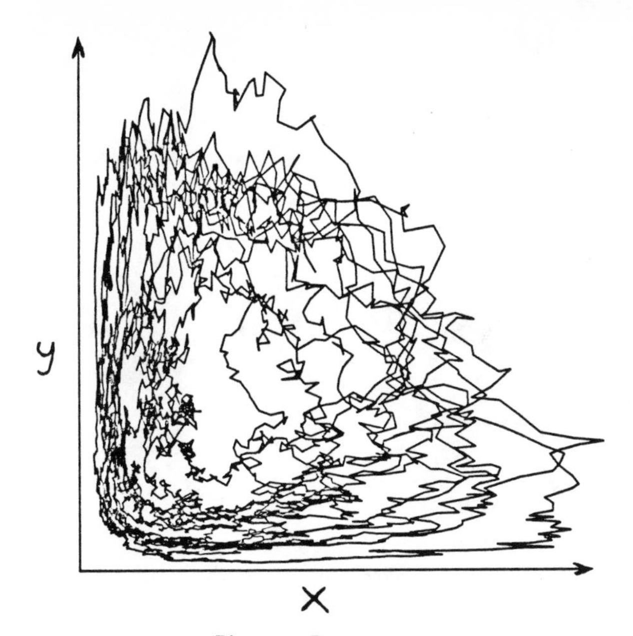

Figure 3
This is a further continuation of Figure 1 up to 45 time units. Additional cycles are indicated as well as a chance excursion of the trajectory near (1, 1) . Of course the trajectory would not remain at (1, 1) , no more than it did in Figure 1.

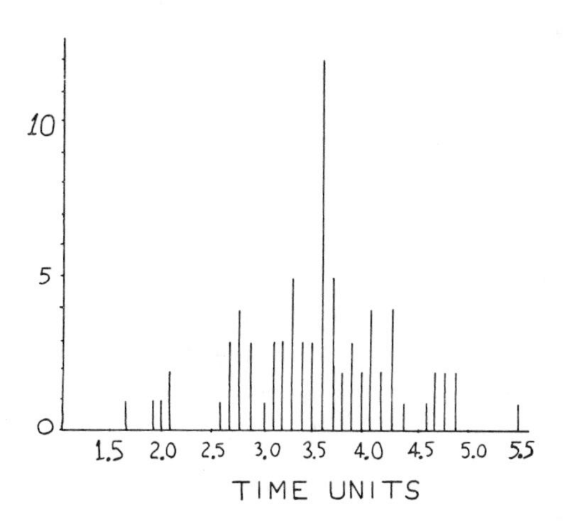

Figure 4
Shown is a histogram of 77 cycles generated using Equations 6a, 6b.
The vertical axis is number of events.

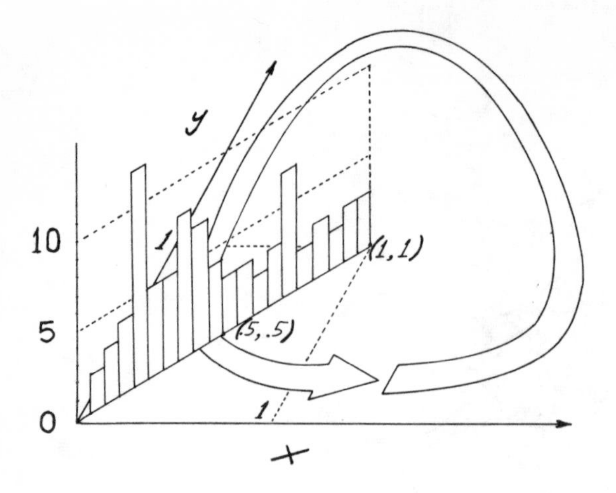

Figure 5
Shown is a histogram of amplitudes of 81 cycles generated using Equations 6a, 6b. The values where the trajectory intersects the $x = y$, $x, y < 1$ line are recorded. The vertical axis is number of events.

for a persistence of regular oscillatory behavior.

The well known Canadian lynx-hare system has been cited by May (1972) as "surely the outcome of some stable limit cycle." Other workers have noted, however, that some data occasionally deviate significantly from a regular cyclic pattern. Numerical data presented in Leigh (1968) show that from 1880 to 1889 a relatively large cycle occurred. In that cycle the lynx population was apparently first to increase, peak, and decline, very much in contrast with our models, especially since the cycle was of large amplitude. Assuming the accuracy of trapping records as sampling, the lynx-hare system remains to be explained, perhaps in terms of a model incorporating additional species.

Finally, we comment on some possible modifications of the model given by Equations 6a, 6b. If smaller perturbations are used, then the populations remain closer to $(1, 1)$, of course. To a certain extent larger perturbations may be used to generate cycles of greater amplitude. However, larger perturbations may lead to a model which is unstable and which may collapse by chance after some (possibly large) number of seemingly stable oscillations. The classical experiment of Huffaker, as described in Leigh (1968), seems to be a straightforward example of such a collapse. Another modification option on Equations 6a, 6b is changing the magnitudes of the interactions. For example, if both species were strongly self-crowding, then the system would be strongly damped. Larger perturbations would be required to send the populations significantly far from (1, 1) , and approximately random, as opposed to cyclic, population fluctuations would be observed.

Multispecific Systems

Near (1, 1) the oscillatory behavior of Equations 6a, 6b is due to the nature of the eigenvalues of the linear approximation matrix, which, due to normalisation, is just the interaction matrix. Any other model with the same first order terms would also exhibit probabilistic limit cycles provided the system were not perturbed out of some sufficiently small neighbourhood of (1, 1) .

In considering the general Lotka-Volterra model for n species

$$dx_i/dt = x_i \left(A_i + \sum_{j=1}^{n} B_i^j x_j\right) \quad (8)$$

there is first the question of the existence of an equilibrium state (for some arithmetic machinery see Strobeck (1973)). However, given the existence of an equilibrium state, we may as well denote it as (1, 1, ... , 1) and consider

$$dx_i/dt = \underbrace{\sum_{j=1}^{n} B_i^j (x_j - 1)}_{\text{linear terms}} + \underbrace{\sum_{j=1}^{n} B_i^j (x_i - 1)(x_j - 1)}_{\text{second order terms}} . \quad (9) .$$

Again the eigenvalues of the linear approximation matrix $\{B_i^j\}$ will characterise the system near equilibrium.

Now it is possible to change the entries in any $n \times n$ matrix less than some fixed but arbitrarily small amount and so obtain a matrix with n distinct eigenvalues. That is, "most" $n \times n$ matrices have n distinct eigenvalues.

Suppose $\{B_i^j\}$ has n distinct eigenvalues. If the system given by Equations (9) is asymptotically stable at equilibrium, then we will observe the following: Real (so negative) eigenvalues will give rise to simple damping to equilibrium, and eigenvalues with an imaginary part will give rise to cyclic patterns.

To clarify this one may, for such a system, resort to a "change of variables." We may call a linear sum of the species x_i a new "species." Biomass can be expressed as a linear sum of species, and biomass is a measure of the amount of a kind of stuff, just as each x_i. For a system with the distinct eigenvalue property, it is always mathematically possible to reexpress the system using a change of variables so that, in first order terms, the system is reduced to independent subsystems, each with one or two "species." The subsystems with one "species" would behave like a prey species in isolation, such as in

Equation (7). The subsystems with two "species" would behave like a predator-prey system, and, if weakly damped, would exhibit probabilistic limit cycles when perturbed. Again all such subsystems using the new coordinates would be completely independent.

If a natural ecosystem could be approximated by Equations (9), then choosing new variables could result in a valuable interpretation of large amounts of interaction data. Systems noted for large natural fluctuations would be of particular interest because of the occurrence of complex eigenvalues and associated subsystems.

Acknowledgments

Thanks are due to Professors Frank Fiala and H. Gray Merriam for many helpful suggestions. Mr. Rick Mallett kindly prepared the computer generated graphs in Figures 1, 2, and 3.

Literature Cited

Felix Albrecht, Harry Gatzke, and Nelson Wax, Stable limit cycles in predator-prey populations, Science 181, (1973) pp. 1073-1074.

M. E. Gilpin, Enriched predator-prey systems: theoretical stability, Science 177, (1972) pp. 902-904.

Egbert R. Leigh, The ecological role of Volterra's equations, Lectures on Mathematics in Life Sciences, American Mathematical Society, (1968) pp. 1-61.

Robert M. May, Limit cycles in predator-prey communities, Science 177, (1972) pp. 900-902.

Robert Rosen, Dynamical System Theory in Biology, Wiley-Interscience, New York, 1970.

M. L. Rosenzweig and R. H. MacArthur, Graphical representation and stability conditions of predator-prey interactions, Americal Naturalist 93, (1963) pp. 209-223.

F. M. Scudo, Vito Volterra and theoretical ecology, Theoretical Population Biology 2, (1971) pp. 1-23.

Curtis Strobeck, N Species competition, Ecology, 54, (1973) pp. 650-654.

Mathematics Department
Carleton University
Ottawa K1S 5B6
Canada

OSCILLATIONS IN BIOCHEMISTRY

Nicholas D. Kazarinoff
Martin Professor of Mathematics
S.U.N.Y. at Buffalo
4246 Ridge Lea Rd.
Buffalo, New York

I first wish to demonstrate an oxidation-reduction reaction involving ordinary inorganic chemicals, a reaction that oscillates, not forever, just for an hour or so. This is the Zaikin-Zhabotinskiǐ reaction whose constituents are water, sulfuric acid, sodium bromate and bromide, malonic acid, and phenanthroline ferrous sulfate. Let me quote Dr. Winfree [8]: "In this aqueous solution phenanthroline catalyzes the oxidation decarboxylation of malonic acid. The reaction oscillates with a period of several minutes, turning from red to blue where phenanthroline is reversibly oxidized. Pseudo waves sweep across the solution (in a petrie dish) at variable speed. In addition, blue waves propagate in concentric rings at fixed velocity from pacemaker centers of varying frequency. Unlike the pseudo waves, these waves are blocked by impermeable barriers. They are not reflected. They annihilate each other when they collide. The outermost wave surrounding a pacemaker is eliminated each time the outside fluid undergoes its spontaneous red-blue-red transition. Because of uniform propagation velocity and mutual annihilation of waves that collide, higher frequency pacemakers control domains which expand at the expense of domains belonging to pacemakers of lower frequency: each slow pacemaker is eventually dominated by the regular arrival of waves at intervals shorter than its period." One can tilt the petrie dish briefly and generate spiral waves! The Zaikin-Zhabotinskiǐ reaction [9,11] is exceedingly complex; and there is dispute between experimental investigators about aspects of the reaction, such as the origin of the circular and the spiral waves and the pacemaker centers.

Spontaneous pacemakers of variable period: this reminds one of Purkinje cells in the human heart. It also calls to mind epidemics. Although the mechanisms of these periodic phenomena are undoubtedly different, mathematically they may have much in common. Currently N. Kopell and L. N. Howard, both mathematicians, are conducting experiments and writing a series of papers on the Zaikin-Zhabotinskiǐ reaction [3,4,5]. They discuss and analyze a nonlinear diffusion model for the self-destroying waves.

A main difference between living and dead is stability. Classically studied chemical reactions, even biochemical ones, are dead - there is a one-point trajectory that is globally asympotically stable. For example, consider a one-substrate (S), one-product (P), enzyme (y) catalyzed reaction involving n catalytic complexes. Here $S, P, y, x_1, \ldots, x_n$ are concentrations and are nonnegative. We may picture the reaction as

$$y + S \underset{k_{-1}}{\overset{k_1}{\rightleftarrows}} x_1 \underset{k_{-2}}{\overset{k_2}{\rightleftarrows}} x_2 \cdots \underset{k_{-n}}{\overset{k_n}{\rightleftarrows}} x_n \underset{k_{-n-1}}{\overset{k_{n+1}}{\rightleftarrows}} y + P$$

The model we use is $(\bullet = d/dt,\ t = \text{time})$:

$$\dot{S} = -k_1 Sy + k_{-1}x_1, \qquad \dot{x}_1 = k_1 Sy - (k_{-1}+k_2)x_1 + k_{-2}x_2,$$

$$\dot{P} = k_{n+1}x_n - k_{-n-1}Py, \qquad \dot{x}_2 = k_2 x_1 - (k_{-2}+k_3)x_2 + k_{-3}x_3,$$

$$\dot{y} = -k_1 yS + k_{-1}x_1 + k_{n+1}x_n - k_{-n-1}Py, \quad \dot{x}_{n-1} = k_{n-1}x_{n-1} - (k_{-n+1}+k_n)x_{n-1} + k_{-n}x_n,$$

$$\dot{x}_n = k_n x_{n-1} - (k_{-n}+k_{n+1})x_n + k_{-n-1}Py.$$

There are two conservation laws:

$$\dot{S} + \dot{P} + \sum_1^n \dot{x}_i = 0 \quad \text{and} \quad \dot{S} + \dot{P} - \dot{y} = 0 .$$

Let $Q^o = (S^o, P^o, y^o, x_1^o, \ldots, x_n^o)$ be the equilibrium state, a point where the right-hand sides of the system are all zero. For each choice of S^o, P^o, there exists one such point if $S^o > 0$ and $P^o > 0$. It is easy to show that the solution Q^o is <u>globally</u> asymptotically stable, namely dead.
Choose as a Liapounov function

$$V = \sum_j z_j \ln \frac{z_j}{z_j^o} - z_j + z_j^o, \text{ where}$$

z_j ranges over each state variable $S, P, y, x_1, \ldots, x_n$.

Note that if $f(z) = z \ln z/z^o - z + z^o$ with $z^o > 0$, then f strictly decreases from z^o at 0 to 0 at z^o, f strictly increases thereafter and the graph of f is concave up. Also note that $-(a-b)\ln {}^a/_b \le 0$ if $a, b > 0$, and that equality holds if and only if $a = b$.

We easily compute $\dot{V}$ to be

$$\dot{S}\ln\frac{S}{S^o} + \dot{P}\ln\frac{P}{P^o} - (\dot{S} + \dot{P} + \dot{x}_2 + \ldots + \dot{x}_n)\ln\frac{x_1}{x_1^o}$$

$$+ \sum_2^n \dot{x}_1 \ln\frac{x_i}{x_1^o} + (\dot{S} + \dot{P})\ln\frac{y}{y^o}$$

or

$$\dot{S} \ln \frac{S y x_1^o}{S^o y^o x_1} + \dot{P} \ln \frac{P y x_n^o}{P^o y^o x_n} + (\dot{P} + \dot{x}_2 + \dots + \dot{x}_n) \ln \frac{x_2 x_1^o}{x_2^o x_1}$$

$$+ (\dot{P} + \dot{x}_3 + \dots + \dot{x}_n) \ln \frac{x_3 x_2^o}{x_3^o x_2} + \dots + (\dot{P} + \dot{x}_n) \ln \frac{x_n x_{n-1}^o}{x_n^o x_{n-1}} .$$

Substitute from the right-hand sides of the system. A typical term is,

(since $k_1 S^o y^o = k_{-1} x_1^o$), $-(k_1 Sy - k_{-1} x_1) \ln \frac{S y x_1^o}{S^o y^o x_1}$ or

$-(k_1 Sy - k_{-1} x_1) \ln \frac{k_1 Sy}{k_{-1} x_1}$, which is nonpositive. It is zero if and only if

$k_1 Sy = k_{-1} x_1$, that is at Q^o. Therefore, $\dot{V} \le 0$, and $\dot{V} = 0$ only at Q^o.

This means we have found a "first octant" Liapounov function; hence we have shown global asymptotic stability of Q^o interior to the region of physical reality. For related biochemical literature see [1,7]. An interesting open question here is: How closely do solutions of the linearized system approximate solutions of the nonlinear system? By the way, one can introduce stronger nonlinearities and still preserve global asymptotic stability of Q^o; namely,

$$\dot{S} = -k_1 S^{\alpha} y^{\gamma} + k_{-1} x_1 , \qquad \dot{x}_1 = k_1 y^{\gamma} S^{\alpha} - k_{-1} x_1 - k_2 x_1 + k_{-2} y^{\gamma} P^{\beta} ,$$

$$\dot{P} = k_2 x_1 - k_{-2} P^{\beta} y^{\gamma} , \qquad \dot{y} = -k_1 y^{\gamma} S^{\alpha} + k_{-1} x_1 + k_2 x_1 - k_2 y^{\gamma} P^{\beta} ,$$

with $\alpha, \beta, \gamma > 0$ and $V = S \ln\left(\frac{S}{S^o}\right)^{\alpha} - \alpha (S - S^o) + \dots$.

There is currently an intensive search for mathematical systems that model lifelike behavior. Ideally, such a system should have an orbitally stable solution and a point of stable equilibrium with the property that the state space can be subdivided into subregions such that every solution originating in a subregion converges to the same solution, either the living, periodic one or the dead, asymptotically stable one. It should also be true that if all (or some) of the nonlinear terms are removed, then the system is "dead." S. Smale has constructed an eight-dimensional nonlinear system with a globally orbitally stable periodic solution, and such that if the interaction terms are zero, then there is a globally asymptotically stable point of equilibrium. Smale's system crudely models a two cell chemical system, each cell containing four chemicals, the cells being separated by a permeable membrane.

L. Glass and S. A. Kauffman [2] have conjectured that the following simpler two cell, two chemical system has the same properties.

$$\dot{x} = \begin{pmatrix} -1-\mu & \mu & 0 & 0 \\ \mu & -1-\mu & 0 & 0 \\ 0 & 0 & -1-\mu & \mu \\ 0 & 0 & \mu & -1-\mu \end{pmatrix} x + \lambda \begin{pmatrix} 1 - S(x_3) \\ 0 \\ 0 \\ S(x_2) \end{pmatrix} .$$

Here λ and μ are positive numbers, and S is a monotone function with an S-shaped graph rising from 0 at 0 to, say, 1 at 1.

It is easy to show that there is a family of equilibrium points of this system, corresponding to choices of S, such that if $\lambda = \kappa\mu$, then for κ near zero the associated linearized system has two conjugate complex roots with negative real parts which increase monotonically with κ and eventually have positive real parts. Thus, by the Hopf bifurcation theorem, the nonlinear system has a periodic solution. To prove orbital stability of that solution in a specific region of the "first octant" appears to be difficult indeed.

I next turn to some models of heart muscle fibres. Zeeman [10] has drawn the following pictures that are associated with the system $\epsilon\dot{x} = -(x^3 + ax + b)$, $\dot{b} = x - x_0$ in which x is fibre length, b is membrane potential, $-a$ is tension and $x_0 > 1/\sqrt{3}$.

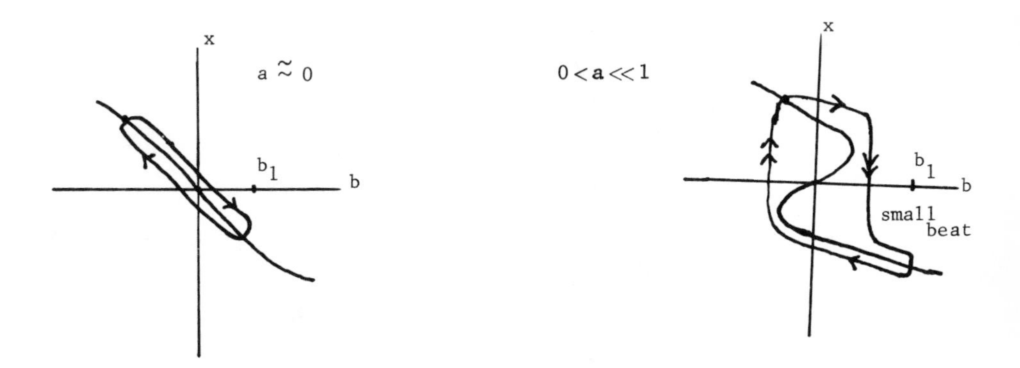

Fig. 1. No tension, bypassed heart

Fig. 2. Atrial fiber

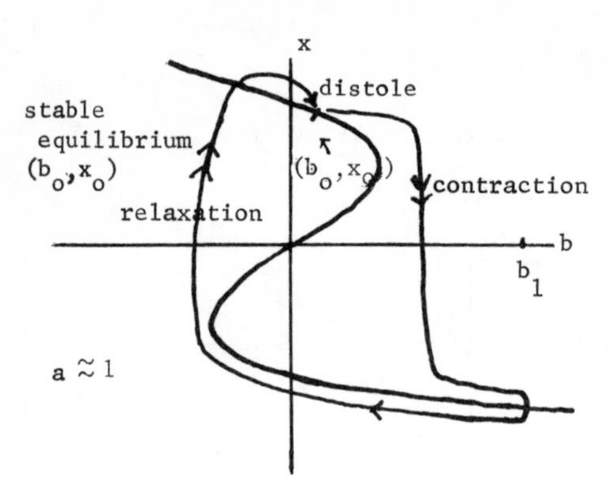

Fig. 3. Normal heart

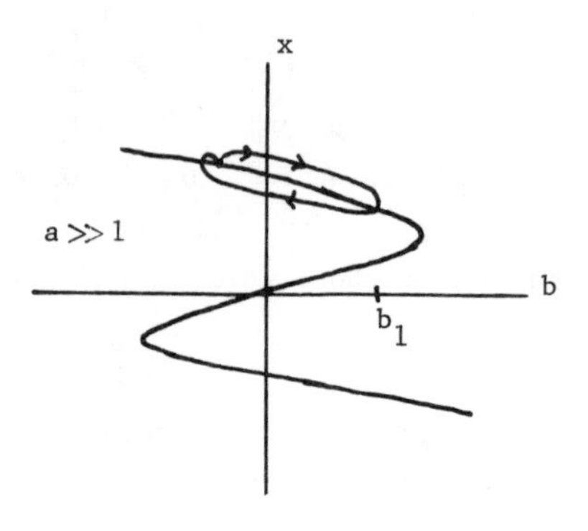

Fig. 4. No beat, too high blood pressure

The appeal of this model is in its dramatic versatility. But to my mind it exhibits a great flaw: there is no modeling of the control mechanism that increases b from b_o to b_1. Such a mechanism is necessary if there are to be closed orbits as illustrated. Krinskiĭ, Pertsov and Reshetilov [6] do have a model with control that exhibits what they call self-exitation or echo. They consider two adjacent "cells" described by the van der Pol-like systems

$$\mu \dot{u}_1 = f(u_1) + v_1 - z(u_1,u_2) \qquad \mu \dot{u}_2 = f(u_2) + v_2 + z(u_1,u_2)$$

$$\dot{v}_1 = \mu \varphi(u_1,v_1) \qquad \dot{v}_2 = \mu \varphi(u_2,v_2)\ ,$$

where $f(u) = u - u^3/3$, $z(u_1,u_2) = 0$ or $\pm k$ as is explained below, and $\varphi(u,v) = -u-bv+a$. Krinskiĭ et al identify u as membrane potential, but they do not identify v. The phase portrait of system (1) with $z = 0$ is indicated in Fig. 5. A trajectory beginning at point 1 leaves the slow manifold at B and converges to stable equilibrium at (u_o, v_o).

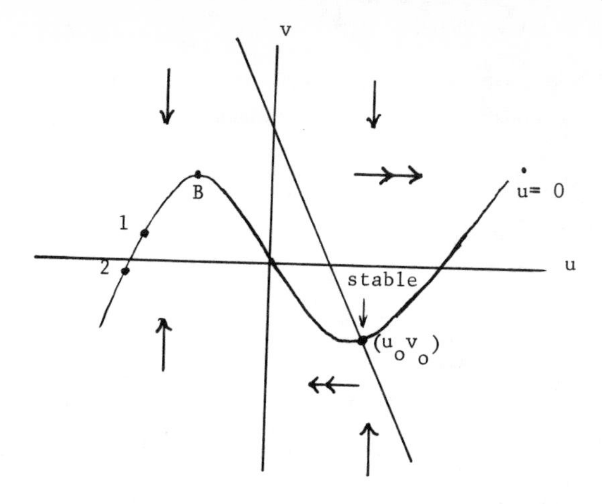

Fig. 5. **Picture when** $z = 0$.

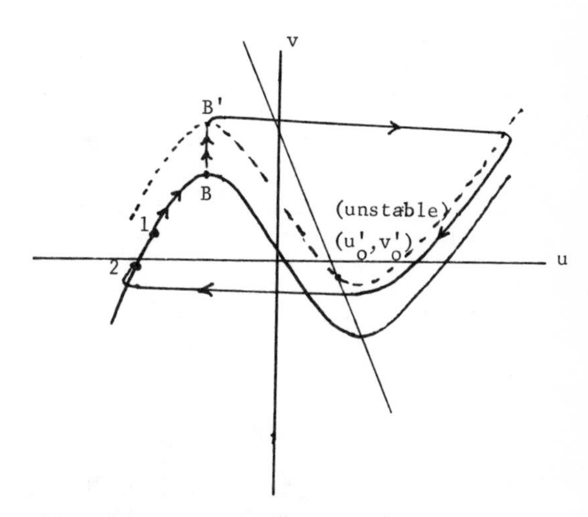

Figure 6. Picture for $z = 0$ and then $+K$

The control function z is described as follows; see Fig. 6 . When the trajectory of a solution of (1) beginning at Point 1 at $t = 0$ reaches B, z switches from 0 and becomes $+K$. The phase point of (1) moves in a cycle as shown: the effect of $z = K > 0$ is that the new equilibrium point (u_o', v_o') is unstable. It reaches the original position of a trajectory of system (2) originating at Point 2 at $t = 0$ just as the phase point of (2) reaches Point 1 (when the phase point of (1) again reaches the slow manifold $z = 0$). The next part of the periodic motion continues according to the rule that when the phase point of (2) reaches B, z becomes $-K$ and cell number (2) beats just as cell number (1) did. For this all to happen φ and f must be chosen properly. Krinskii et al have no proof that it does, only numerical evidence.

These investigations give a glimpse of a yet to be written chapter in the theory of ordinary differential equations, one whose content will be exciting to mathematicians. It is to be hoped that the theorems discovered will be general enough in scope and specific enough in their qualitative and quantitative feature to give biological scientists new insights into biological and biochemical systems, even suggesting important new experiments.

References

1. Boyer, P. D., Editor, The Enzymes, vol. II., Kinetics and Mechanism, Chapter 2, Academic Press, N. Y. 1970.

2. Glass, L. A. and Kauffman, S. A., Co-operative components, spatial localization and oscillatory cellular dynamics, J. Theoretic. Biology 34(1972), 219-237.

3. Howard, L. N. and Kopell, N., Spatial structure in the Belousov reaction I. External gradients; II. Diffusion and target patterns. Preprints.

4. Kopell, N. and Howard, L. N., Horizontal bands in the Belousov reaction, Science 180(1973), 1171.

5. ————————————, Plane wave solutions to reaction diffusion equations. Preprint.

6. Krinskiĭ, V. I., Pertsov, A. M. and Reshetilov, A. N., Investigation of one mechanism of origin of the ectoptic focus in modified Hodgkin-Huxley equations, Biofizika 17(1972) No. 2, 271-277.

7. Walter, C., Enzyme Kinetics, Ronald Press, New York 1966.

8. Winfree, A. T., Spiral waves of chemical activity, Science 175(1972), 634-636.

9. Zaikin, A. N. and Zhabotinskiĭ, A. M., Concentration wave propagation in a two-dimensional liquid-phase self-oscillating system, Nature 225(1970), 535.

10. Zeeman, E. C., Differential equations for the heart and nerve . Preprint.

11. Zhabotinskiĭ, A. M., Dokl. Akad. Nauk. S.S.S.R. 157(1964), 392.

A SIMPLE TEST FOR TAG-RECAPTURE ESTIMATION

William Knight

Mathematics Department
University of New Brunswick

ABSTRACT

The assumptions upon which tag-recapture methods for estimating population size or exploitation rate are based are ever a source of doubt. A simple check of these assumptions in the context of fisheries population dynamics is proposed. This check, like any test of scientific hypotheses, disproves the assumptions if it fails; and supports but does not prove the assumptions if it succeeds. The check is tied to practical requirements. Failure of this check is not uncommon and may even be the usual state of affairs.

ARE SPECIES ASSOCIATION COEFFICIENTS REALLY NECESSARY?

William Knight

Department of Mathematics
University of New Brunswick
Fredericton, N.B.

Albert V. Tyler

Fisheries Research Board of Canada
Biological Station
St. Andrews, N.B.

ABSTRACT

This paper advocates the thesis that species association coefficients are to be used as little as possible, and if they must be used, should be eliminated after use. As a preliminary, the all too obvious flaws of such coefficients are reviewed. Trying to trace ecological meaning back through association coefficients is not tempting in view of the large numbers of coefficients generated in a species assemblage study, and in view of the dubious interpretation of some coefficients. Further, the technique of first grouping or ordinating species, and then attempting to relate the arrangements to obvious environmental features, is an indirect approach that should be deferred in favour of methods relating the presence of species directly to those features. Examples are mentioned to show how association coefficients can be eliminated, at both the verbal and mathematical levels, and to show how environmental features can be brought directly into the analysis.

A MODEL OF MORPHOGENESIS

Simon Kochen

ABSTRACT

Department of Mathematics Princeton University

Turing in 1952 proposed a mathematical model in which morphogenetic change occurs through the onset of thermodynamic instability. The major mechanism of transport in the model was diffusion. This led to a system of non-linear partial differential equations

$$\frac{\partial x}{\partial t} = \mu \nabla^2 x + Vx \tag{1}$$

where $x(\bar{r},t)$ denotes the vector of change of chemical concentrations from an equilibrium point, $V(x,\bar{r},t)$ is the reaction rate in the cell at position $\bar{r}$, and μ is the diffusion constant. This was applied to a one-dimensional homogeneous ring to show how symmetry will naturally break down by the onset of instability.

We apply the model (1) to the 3-dimensional case of a fertilized egg. Starting with a homogeneous solid spherical morula, we find that the formation of the hollow blastula and its invagination occur naturally in the model, as instability is reached with changing parameter of diffusion due to cell cleavage. However, the breaking of spherical symmetry via invagination occurs in this model along a randomly determined axis, which does not conform to experimental evidence. This led us to investigate the process when a lower degree of symmetry (e.g. radial or bilateral) is assumed initially for the fertilized egg, as is experimentally the case. Here exact solutions are not possible and perturbation methods (of Rayleigh-Schrödinger) were applied. In this case the lower symmetry was not broken. In particular the axis of invagination was determined by an original axis of symmetry. This suggests the explanation in this model for the prevalence of radially and bilaterally symmetric multi-cellular organisms.

For better conceptual understanding and to deal with more complicated cases we have also used group theoretic methods. These methods yield "selection rules" for the preservation (or breaking) of symmetries of solutions to (1) without having to give detailed solutions. This approach is analogous to the group-theoretic derivation of selection rules in quantum mechanics for the corresponding Schrödinger equation.

FLAGELLAR GROWTH

Elinor Miller Levy
Department of Pathology
University of British Columbia
Vancouver, B.C.

The flagellum is the tail-like locomotive organelle of the cell. Both complex and bacterial flagella grow by the polymerization of protein precursors at the distal tip of their flagellar fibers; that is they grow from the end furthest removed from the cell body (Rosenbaum and Child, 1967, Tamm, 1967, Iino, 1969). Generally the rate of flagellar growth is fastest initially and decreases continuously until growth stops at a species specific length. If a flagellum is broken, it will regenerate. The kinetics of regenerative growth parallel those of normal growth with the initial rate of elongation being a decreasing function of the stump length (figure 1). Both the physical process and kinetics of elongation thus suggest that the growth rate and final length of the flagellum are controlled by a decreasing concentration of a polymerization participant at the building site. In this paper a model based on this hypothesis is discussed.

Let us consider growth in a group of flagellar fibers. It is assumed for now that the controlling factor is a protein building unit. For simplicity it is assumed that the concentration of this precursor is kept constant at the flagellar base and that its molecules move to the tip by diffusion along the flagellum. The concentration at the tip is assumed to decrease linearly with flagellar length; i.e., the concentration at the tip for fibers having j subunits, p_j, is given by $p_j = p_o(1 - \frac{j}{m})$, where p_o is the concentration at the base and m is the maximum number of subunits per fiber.

This proposed gradient might arise, for instance, from the leakage of precursor molecules through the flagellar tip (Levy, 1973a). It can be seen that this model will result in a decreasing growth rate with increasing flagellar length without the need for a decreasing supply of the controlling factor within the cell body. Such a spatial rather than temporal change in concentration is necessitated by the regenerative ability of flagella.

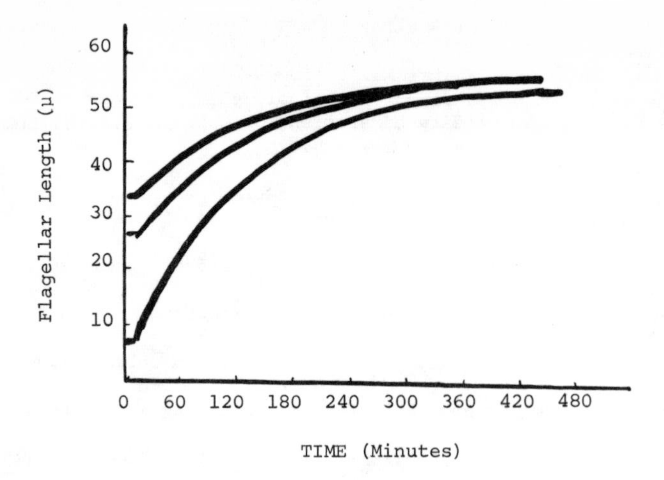

Figure 1. Growth curves of Peranema flagella of varying initial lengths. From S. Tamm, J. Exptl. Zool. 164, p.173.

Let f_o be the number of flagellar bases, f_j the number of fibers having j subunits, and p_j the concentration of building units at the reaction site of such flagella fibers; then kinetically the polymerization process can be approximated by the following set of equations:

$$\frac{df_o}{dt} = -kf_o p_o$$

$$\frac{df_j}{dt} = -kf_j p_j + kf_{j-1}\, p_{j-1} \qquad \text{for } j \neq 0.$$

These can be solved using Laplace transforms. If f is the total number of fibers in the system under consideration and growth is starting from the base, f_j is given by:

$$f_j = f \frac{m!}{(m-j)!j!} \; e^{-p_o kt} \sum_{i=o}^{j} \frac{(-1)^{j-i} j!}{(j-i)!\; i!} \; e^{ip_o kt/m} . \tag{1}$$

The fraction of fibers having j units at any time, $\frac{f_j}{f}$, can be treated as a probability distribution. The average number of units per fiber is found as:

$$\bar{\chi} = m(1-e^{-p_o kt/m}) . \tag{2}$$

Using a similar argument for the regenerative case in which growth starts from a flagellum having n units, the average fiber length is:

$$\bar{\chi} = n + (m-n)\;(1-e^{-p_o kT/m}) . \tag{3}$$

The two expressions for average fiber length in normal and regenerative growth, (2) and (3), are seen to be equivalent when the total growing time, t, is separated into the average time spent growing to length n and the growing time thereafter, T. Thus this model suggests that normal and regenerative growth curves can be superimposed. This is in agreement with Tamm's data, part of which is seen in figure 1.

Although we have assumed up until now that the controlling factor is a fiber protein precursor, our model would work equally well for a smaller controlling molecule such as an ion or nucleotide. In this case the small molecule would be assumed to participate but not necessarily be consumed in the polymerization reaction. Its concentration would decrease linearly with reaction site distance, while that of the protein subunits would now remain constant. The schematic representation of the reaction would be:

$$f_j + p_o + u_j \xrightarrow{k} f_{j-1} + (u_i) ,$$

while the average number of subunits would be given by:

$$\bar{\chi} = m \, (1-e^{-u_o p_o kt/m}) ,$$

which is obviously formally the same as equation (2).

Further consequences of this model can be elucidated by the use of coupled differential equations for the controlling factor's concentration and flagellar length (Levy, 1973b). Although this model is greatly simplified, the major features of flagellar growth can be understood in terms of it.

REFERENCES

Iino, T. *Bact. Rev.* 33, 543 (1969).
Levy, E.M. In preparation (1973a).
Levy, E.M. *J. Theoret. Biol.* In press (]973b).
Rosenbaum, J. and Child, F. *J. Cell Sci.* 34, 345 (1967).
Tamm, S. *J. Exptl. Zool.* 164, 163 (1967).

SOME REMARKS ON RANDOM SETS MOSAICS

by

J.E. Lewis and Thomas Rogers*

Mathematics Department
University of Alberta

1. INTRODUCTION

Given a set of points S having no accumulation points contained in a subset T of the plane, associate with each element a of S the subset S_a of points in T closer to a than the other points of S. The subsets S_a are convex polygons and partition T. T is not necessarily closed or bounded, but of course, if T is compact then S is finite. These polygons are variously referred to as Dirichlet cells, Meijering cells (1953), Voronoi regions (1908), Gilbert cells (1961), cells or tiles, and the partition is called an S-mosaic or mosaic. If the elements of S are sprinkled randomly in T call the partition a random S-mosaic. The point a is called an S-point for the tile S_a. Generalizations of this idea to higher dimensional spaces (even to metric spaces) are apparent. We have not looked into any of these in detail, although metric space (using just metric convexity?) or normed linear space generalizations of known E_n results might be interesting.

C.A. Rogers in his book Packing and Covering (1964) discusses Voronoi polyhedra in his chapter which states that "packings of spheres cannot be very dense." He discusses the disection into simplicies of these polyhedra in E_n. E.N. Gilbert in 1961 considered random S-mosaics and their generalization to

---

* The research of this author was supported by NRC Grant No. 5210.

Johnson-Mehl mosaics (1939) in which the S-points now have a "time of arrival" and the tiles are star-shaped. S-mosaics are generally "attractive" tessellations, resembling mud-cracks or dragonfly wings or many other of the mosaics found in nature, as exemplified, for example, in D'Arcy Thompson's well-known treatise (1942). We will mention some specific biological references later.

2. QUESTIONS ON S-MOSAICS

(a) Let n points be sprinkled in a convex set T and let the n tiles of the resulting S-mosaic be ordered according to area: $S_1 \leq S_2 \leq \ldots \leq S_n$. Then what is the expected area (over the original point distribution) of the S_i's ? This suggested partial generalization of MacArthur's broken stick model, might be referred to as the broken dish model.

(b) If a convex set is polytopically subdivided, find conditions that the subdivision is an S-mosaic. (On the line, a necessary but not sufficient, condition is that the length of each interval is less than the sum of the adjacent intervals.)

(c) Using the vertices of the S-tiles as new points to generate a different S-mosaic, what happens as this process is continued indefinitely? For the line, it seems as though the tiles become of equal length.

(d) What if the collection S is replaced by sets rather than points? The resulting tiles are no longer convex polytopes.

(e) S-points are not unique. However, they appear to be unique in E_3.

(f) Find a NASC that a broken stick be an S-mosaic. See (c).

(g) For the trivial case that T = the disc in the plane with radius a and S is a random uniform sprinkling of two points in T, the expected size of the smaller segment is

$$E(A) = \frac{2}{(\pi a^2)^2} \int_{r_2=0}^{a} \int_{r_1=0}^{r_2} \int_{\theta_1=0}^{2\pi} \int_{\theta_1=0}^{2\pi} A(\ell) r_1 dr_1 d\theta_1 r_2 dr_2 d\theta_2$$

where $0 \leq r_1 \leq r_2 \leq a$, $\ell = \dfrac{r_2^2 - r_1^2}{\sqrt{r_1^2 + r_2^2 - 2r_1 r_2 \cos(\theta_1 - \theta_2)}}$

and $A(\ell) = \dfrac{\pi a^2}{2} - \ell\sqrt{a^2 - \ell^2} - a^2 \arcsin\left(\dfrac{\ell}{a}\right)$.

We obtained, with the help of J. Mosevich, M. Marsden and I.B.M.,

$$E(A) = \frac{8}{\pi}\left(\frac{2a}{3}\right)^2 \qquad \text{(conjecture)}.$$

Kendal and Moran's monograph Geometrical Probability (1963) is a good reference for problems like this.

3. APPLICATIONS TO BIOLOGY

(a) In Bendell's and Elliott's blue grouse monograph (1967) there are slightly skewed S-mosaics in their ecological maps. The tiles are territories of the males. There is an S-mosaic in the husband-wife territories of pikas, as in the 1973 Ph.D. thesis written by Stephen Tapper under Fred Zwickel at Alberta. In general, we suggest that the concept of S-mosaic might be extremely useful in *any* study of behaviour, territory or competition. A wonderful example is found in Konrad Lorenz' King Solomon's Ring (1952, p. 26) in which the territorial behaviour of Stickleback perch is described. A humorous example is found in Mario Puzo's The Godfather Papers wherein an encounter between Frank Sinatra and John Wayne affords an almost perfect mathematical definition of an S-mosaic.

(b) E.C. Pielou discusses S-mosaics and contrasts them with "lines mosaics" in her recent textbook An Introduction to Mathematical Ecology. She is mostly interested in using mosaics in the context of sampling and refers to a 144 page paper by B. Matérn (1960) dealing with their use in forest surveys.

(c) We suggest the use of S-mosaics in competition studies, where spatial patterns are considered. For example, in parasitology the territories of intestinal helminths might be considered with S-mosaics in mind.

(d) _If_ contact inhibition occurs among root balls of plants, S-mosaics will result if the area is colonized by seedlings of the same age.

(e) If the above 2-D example is extended to Hutchinson's idea of a vector-niche in E_n, then more than just space variables can be included. For example, time, temperature, and other environmental factors might be considered. In passing, we would like to mention that it also might prove profitable to use the Hutchinson-MacArthur vector-niche concept in conjunction with an information theoretic analysis of species-niche matrices; note the matrix entries could still be numbers. We are trying to convince Don Gallup, a limnologist at Alberta, to use this idea in a certain aquatic plant and animal sampling program of his. Note the Shannon-Weaver formula or Simpson's index could be used without change.

(f) Arthur Winfree in a sequence of papers on spiral activity in chemical reactions describes how several spiral wave fronts produce an S-mosaic, essentially because the wave fronts are self-annihlating on contact. See his papers (1972, 1973) for detail and possible applications to fungoid growth patterns and social amoebas.

(g) John Maynard Smith in his new book _Models in Ecology_ discusses S-mosaics, although they are not identified as such (Chapter XII, Territorial Behavior). He also discusses a variation wherein once an S-mosaic is established, the _centroids_ of the tiles (not necessarily S-points) generate a new S-mosaic, this process being continued indefinitely. He claims the limiting configuration is a hexagonal tessellation. Maynard Smith cites Kreb's work with the Great Tit, _Parsus major_, as providing evidence for S-mosaic behavior. The problem of the territorial configuration due to nonsynchronous arrival times of nesting pairs might be handled by the previously mentioned Johnson-Mehl mosaic or "raindrop model".

(h) A. R. Fraser and P. van den Driessche (1972) in a sampling study consider a tessellation into triangles of a finite set with application to neighbour pairs. M. Matern remarks in a discussion concluding the paper, that such triangles seem to be a specialization of the S-mosaic polygon idea.

4. FURTHER CONSIDERATIONS

The concept of an S-mosaic might be extended to include what we shall call epsilon-inhibitory and epsilon-tolerant mosaics; here all the tiles are uniformly separated by a distance ε , or dually overlap by an amount ε . This idea might also be modified to allow the ε-factor to vary locally. Again we suggest this idea might prove to have important applications to behaviourism, competition, or any "contact inhibition" situation.

5. SUMMARY AND CONCLUSIONS

S-mosaics with their constituting tiles seem to be natural ecological territories to study. Each tile consists of all points in the area closer to given point, taken from a set S , than to all the other points in S . Hence the points in the tile, this "closest distance" set, "belong" to their corresponding S-point, and so the apparent applications to problems in competition. Epsilon-inhibitory or tolerant mosaics might allow, for example, for different levels of aggression or maturity or poisoning, and so on. We contrast this direct use of mosaics as opposed to their statistical use as in sampling studies.

REFERENCES

[1]. Bendell, J.F. and Elliott, P.W. Behaviour and the regulation of numbers of blue grouse. Canadian Wildlife Service Report Series, No. 4, Ottawa (1967).

[2]. Gilbert, E.N. Random subdivision of space into crystals. Ann. Math. Statist. 33, 958-972 (1962).

[3]. Fraser, A.R. and van den Driessche, P. Triangles, density, and pattern in point populations. 3rd Conference of Advisory Group of Forest Statisticians. I.U.F.R.O. 277-286 (1972).

[4]. Johnson, W.A. and Mehl, R.F. Reaction kinetics in processes of nucleation and growth. Trans. A.I.M.M.E. 135, 416-458 (1939).

[5]. Kendall, M.G. and Moran, P.A.P. Geometrical Probability. Charles Griffin and Company (1963).

[6]. Lorenz, K.Z. King Solomon's Ring. Mathean Company (1952).

[7]. Matérn, B. Spatial variation stochastic models and their application to some problems in forest surveys and other sampling investigations. Medd. fran states Skogsforskringsinstitut 49, 1-144 (1960).

[8]. Meijering, J.L. Interface area, edge length, and number of vertices in crystal aggregates with random neucleation. Phillips Res. Rep. 8, 270-290 (1953).

[9]. Pielou, E.C. An Introduction to Mathematical Ecology. Wiley-Interscience (1969).

[10]. Puzo, Mario The Godfather Papers. Fawcett Crest Paperback (1972).

[11]. Rogers, C.A. Packing and Covering. Cambridge University Press (1964).

[12]. Smith, J. Maynard Models in Ecology. Cambridge University Press. (In press). (1973).

[13]. Tapper, S. The spatial organization of pikas [Ochotona], and its effect on population recruitment. Ph.D. thesis, Department of Zoology, University of Alberta. 154 pages (1973).

[14]. Thompson, D'Arcy W. On Growth and Form. 2nd Ed. Vol. 2, Cambridge University Press (1942).

[15]. Winfree, A.T. Spiral Waves of Chemical Activity. Science, 175, 634-636 (1972).

[16]. Winfree, A.T. Spatial and temporal patterns in the Zhabotinsky reaction, Presented at the Aharon Katchalsky Memorial Symposium, Berkeley, California (1973).

QUALITATIVE BEHAVIOR OF STOCHASTIC EPIDEMICS

Donald Ludwig

New York University

251 Mercer Street

New York, New York

Epidemic theory suffers from a great disproportion between the number of parameters which are needed to describe the important phenomena, and the number of parameters which can be estimated from the data. In order to understand the behavior of epidemics, one should take account of effects such as latency periods, variations in infectivity from one individual to another, and complicated mixing structures in the population. On the other hand, complete and reliable data are difficult to obtain. Often, one can at most hope for data on the final size of the epidemic, i.e. the total number of individuals who were infected at some time or other in the history of the epidemic.

Such a situation might be considered to be unfavorable for mathematical or theoretical investigation. However, great simplifications can be made if attention is restricted to effects which alter the final size distribution.

As a simple illustration, consider a population which consists of a single infective and a single susceptible. Let $\beta(t)$ be the infectivity of the infective, as a function of the time elapsed since his own infection. Then the probability of infectious contact between the two individuals in the interval $(t, t + \delta t)$ is given by $\beta(t)\,\delta t + o(\delta t)$. Let $q(t)$ denote the probability that there is no such contact before epoch t . Then

$$q(t + \delta t) = q(t)\,(1 - \beta(t)\,\delta t + o(\delta t)) \,, \tag{1}$$

and hence q satisfies

$$\frac{dq}{dt} = -\beta(t)\,q(t) \,. \tag{2}$$

The solution of (2) which assumes the value 1 at $t = 0$ is

$$q(t) = \exp \left[- \int_0^t \beta(\tau)\, d\tau \right] . \tag{3}$$

In particular,

$$q(\infty) = \exp \left[- \int_0^\infty \beta(\tau)\, d\tau \right] . \tag{4}$$

Although a complete knowledge of $\beta(t)$ is necessary in order to specify q for all t, we see that $q(\infty)$ depends only upon the integral of β. Since $q(\infty)$ is the probability that the "epidemic" will terminate with one susceptible left, we see that the distribution of final sizes depends only upon the integral of β. Latency effects, for example, do not influence the final size distribution. Similar results can be proved for quite general epidemic models, with arbitrary initial numbers of infectives and susceptibles. If the infection is spread through a sequence of single infectious contacts between infectives and susceptibles, then the final size distribution of the epidemic depends only upon the integral of the infectivity of an infective. This result is proved for models with several types of infectives or susceptibles in Ludwig (1973).

A consequence of this result is that, for the purpose of computing final size distributions, each epidemic process can be replaced by a Markov chain. One simply concentrates the infectious period of each infective into an instant which is one time unit after his own infection. It is relatively easy to compute final size distributions for such simplified models.

The qualitative behavior of final size distributions is important for any comparison of theory with the data. Two important questions are (1) What is the probability of a major outbreak? (2) If a large epidemic does occur, what proportion of the population will be affected? Approximate answers to these questions are given in Ludwig (1974), for a large class of stochastic epidemics. In addition, a method is provided to compute approximate final size distributions. The main features of this distribution depend upon only two parameters which provide the answers to the questions posed above.

If the total population is large, but the number of infectives is small, then the epidemic process can be approximated by a branching process. Let ρ_0 denote the extinction probability for the branching process. The proportion of epidemics with small final sizes is approximately ρ_o. Therefore $1 - \rho_o$ provides an answer to the first question.

If the number of infectives is large, then the epidemic process is approximated by a "quasi-deterministic" process, where the numbers of infectives and susceptibles have a joint normal distribution. The expected numbers of infectives and susceptibles satisfy deterministic equations. Let S_∞ denote the expected final number of susceptibles. This parameter provides an approximate answer to the second question.

The parameters ρ_o and S_∞ may each be determined by solving a transcendental equation. More detailed information is provided by the final size distribution for the approximating branching process, and by computing the variance of the number of susceptibles for large time. This information can be obtained from simple recursion formulas.

Therefore, by combining the results of Ludwig (1973) and (1974), the qualitative behavior of general epidemic models can be described in terms of just two parameters.

This research was supported in part by the National Science Foundation under Grant Number GP 32996X2 .

References

1. Ludwig (1973) Final size distributions for epidemics. To appear in Math. Biosciences.

2. Ludwig (1974) Qualitative behavior of stochastic epidemics. Submitted to Math. Biosciences.

STOCHASTIC MODELS FOR CELL PROLIFERATION

by Peter D. M. Macdonald

Department of Applied Mathematics
McMaster University, Hamilton, Ontario

INTRODUCTION

The growth and form of any biological tissue are determined by the division rates of proliferating cells, the dynamic relationship between proliferating and quiescent cells and the growth of the cells themselves. This paper develops a model for cell proliferation which is valid when a cell population has experienced relatively constant conditions for a number of generations, with generation times variable but successive generation times uncorrelated. The object is to show how modeling has led to methods for analysing data from thymidine-labelling experiments of the sort that are being done by biologists interested in the proliferative characteristics of cells, especially in studies of tumour growth and the development of meristematic tissue in plants.

This paper describes one such experiment, the "fraction labelled mitoses" method, abbreviated flm[1], where tritium-labelled thymidine (a specific precursor of DNA) is made available to the cell population for a brief period of time and the fraction of mitotic cells (cells in the visible act of division) which carry the label is determined at successive later times. The fraction of mitoses which are labelled, plotted as a function of time, is called the fraction labelled mitoses curve. In the past fifteen years a great many biologists have published the results of such experiments and a number of mathematicians have published theoretical derivations of the flm curve (see, for example, Macdonald, 1970, or Barlow and Macdonald, 1973, for references). The theoretical derivations have tended to be complicated and difficult to apply to the analysis of actual experimental data. The biologists have to a large extent ignored the theoretical developments and many continue to analyse their data in ways that are at best *ad hoc* and at worst have no logical foundation. When analyses have been used to compare different biological material or work done at different laboratories, basic statistical questions such as the precision of the analysis have been overlooked.

I shall give what I feel is a particularly simple and general development of the theory for an asynchronous cell population in a steady state of exponential growth and show how Dr. Peter Barlow and I have used this to map out the proliferative characteristics of cells in the root meristem of *Zea mays*, c.v. Golden Bantam.

---

[1] Also called "percent labelled mitoses" by some authors and abbreviated PLM, FLM, flm or f.l.m. at the whim of various journals.

Finally, I shall indicate how more appropriate models for the root meristem might be developed.

A MODEL FOR CELL PROLIFERATION

It has been established that many kinds of cells pass through a sequence of four distinct phases during the course of a mitotic cycle. The phases are called G_1, S, G_2 and M, where S is a period of DNA synthesis distinguished by the uptake of tritiated thymidine and M is mitosis, the visible act of division. We assume that the durations follow some joint distribution with a probability density function $\phi_{1,2,3,4}(u_1,u_2,u_3,u_4)$ and a Laplace transform denoted by $\phi^*_{1,2,3,4}(s_1,s_2,s_3,s_4)$. The mean durations of these phases are typically the parameters of interest. The observed variability in apparent phase duration will be partly a measurement threshold artifact if the transitions into and out of S are not sharply defined. In fact, if S starts more abruptly than it stops the duration of G_2 may appear more variable than the duration of the combined phases $S+G_2$, giving the impression that S and G_2 are negatively correlated. It is therefore essential that a model allow for possible correlations between phases. However, variances and correlations estimated from experimental data may have no simple biological interpretation.

There is some disagreement as to what is meant by a "nonproliferating" or "quiescent" cell. Organized cell populations are usually made up of cells which divide at more or less regular intervals plus a number of cells which have not necessarily lost the ability to divide but show no inclination to do so. These latter cells are sometimes said to have entered a G_0 phase from which they might at some later time re-enter normal mitotic cycles. The way in which cells are selectively "switched on and off" according to the overall requirements of the tissue is not properly understood.

The growth fraction, denoted by GF, is defined as the proportion of cells in a population which are actively proliferating at a given point in time. While it may be safe to assume that all cells in S or M are proliferating cells, there is no way to determine which of the G_1 and G_2 cells have been "switched off" into quiescence. Furthermore, quiescent cells which re-enter the cycle will appear not as quiescent cells but as proliferating cells with very long cycle times. It is thus a problem, in any attempt to estimate the growth fraction, to decide just what has been measured (Clowes, 1971).

Each mitosis results in two daughter cells; however, we let $A \leq 2$ denote the mean number of daughters to remain proliferative after a mitosis. Assuming only that $A>1$, that successive cycles are independent and that the proliferative characteristics of the population are not changing with time, cell proliferation is described by an age-dependent branching process and well-known results are available to describe the ultimate age structure of the cell population (Harris, 1963, Chapter VI). These results have been extended by Macdonald (1973) and Brockwell and Kuo (1972) to give the limiting age and phase structure of a multiphase branching process and the relevant results are summarized here.

The notation uses the subscripts 1, 2, 3, 4 to denote the phases G_1, S, G_2, M, respectively, and combines subscripts to denote combined phases; the subscript 23, for instance, refers to S and G_2 combined and regarded as a single phase. The complete cycle is indicated by the absence of a subscript. A * denotes a Laplace transform.

The basic formula, derived by Macdonald (1973), gives the joint distribution of the phase of a randomly chosen cell together with the durations of all past, present and future phases. For example, the probability that a randomly chosen cell is in phase 3, having spent times u_1, u_2 in the previous phases and attained age x in phase 3, and is to remain in phase 3 for an additional time y and then spend time u_4 in phase 4, is given by the expression

$$\frac{A}{A-1} k e^{-k(u_1+u_2+x)} \phi_{1,2,3,4}(u_1,u_2,x+y,u_4)\, du_1 du_2 dx dy du_4 \tag{1}$$

where k is the positive root of

$$A\, \phi^*(k) = 1 \quad , \tag{2}$$

$\phi^*(s)$ being the Laplace transform of $\phi(u)$, the probability density function for cycle duration. All the following distributions are deduced from (1) or from similar expressions.

The duration of phase 3, for a cell sampled just as it <u>enters</u> phase 3, has probability density function

$$h_3(u) = \int_0^\infty e^{-kv} \phi_{12,3}(v,u)\, dv \;/\; \phi^*_{12}(k) \tag{3}$$

and cumulative distribution function

$$H_3(u) = \int_0^u h_3(v)\, dv . \tag{4}$$

If phase 3 is not independent of previous phases then $h_3(u)$ will differ from $\phi_3(u)$, which is the probability density function for the duration of phase 3 for a cell sampled just as it <u>enters phase 1</u>. It should be clear, however, that the distributions $h_{12}(u)$ and $\phi_{12}(u)$ are identical, as are $h_1(u)$ and $\phi_1(u)$ or $h(u)$ and $\phi(u)$. It can also be shown that

$$h^*(k) = h_1^*(k)\ h_2^*(k)\ h_3^*(k)\ h_4^*(k) \tag{5}$$

and, in fact, the correlated-phase model with joint distribution $\phi_{1,2,3,4}$ is in certain respects equivalent to an independent-phase model with the distributions of phase duration given by the appropriate h_i's, rather than by the marginal distributions of $\phi_{1,2,3,4}$.

The duration of phase 3, for a cell sampled just as it leaves phase 3, has probability density function

$$f_3(u) = e^{-ku}\ h_3(u)\ /\ h^*{}_3(k) \tag{6}$$

and cumulative distribution function

$$F_3(u) = \int_0^u f_3(v)dv \quad . \tag{7}$$

Relationships between the ϕ's, h's and f's, both in general and in terms of standard distributions, are discussed in Macdonald (1973).

The fractions of the proliferative population in phases 2 and 4 at any given time are

$$P_2 = \{A/(A-1)\}\ h_1^*(k)\ \{1 - h_2^*(k)\} \tag{8}$$

and

$$P_4 = \{A/(A-1)\}\ h_{123}^*(k)\ \{1 - h_4^*(k)\}, \tag{9}$$

respectively.

Finally, let $\psi_{23(4)}$ denote the probability density function for age, measured from the start of phase 2, of a cell sampled from phase 4. It can be shown that

$$\psi_{23(4)}(t) = ke^{-kt}\ \frac{H_{23}(t) - H_{234}(t)}{h_{23}^*(k) - h_{234}^*(k)} \tag{10}$$

with cumulative distribution function

$$\Psi_{23(4)}(t) = \{D_{23}(t) - h_4^*(k)D_{234}(t)\}/\{1 - h_4^*(k)\}, \tag{11}$$

where

$$D_i(t) = F_i(t) - e^{-kt}H_i(t)\ /\ h_i^*(k) \ .$$

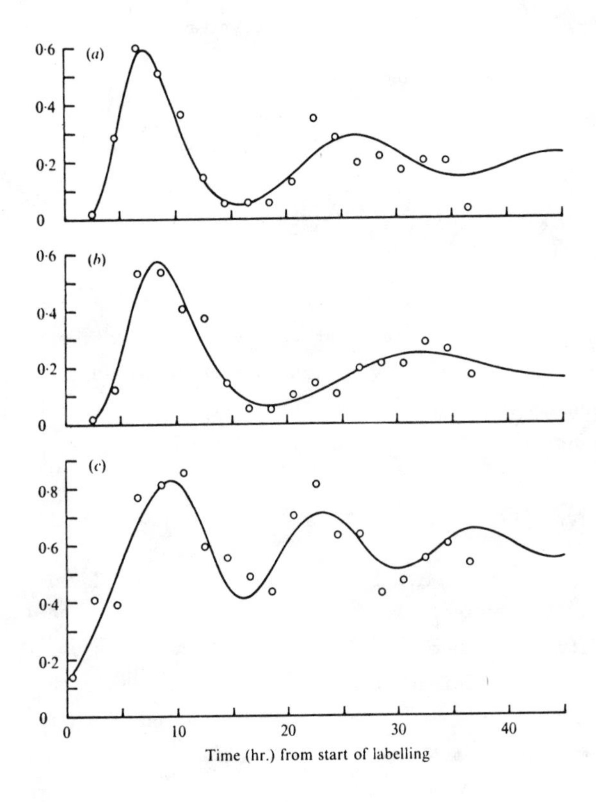

Figure 1. Fraction labelled mitoses curves fitted to data from the root meristem of Zea mays.
(a) Cortex at 700 µm; (b) cortex at 400 µm;
(c) cap columella initials.

THE ANLYSIS OF PULSE-LABELLING EXPERIMENTS

Suppose that tritiated thymidine is made available to the cells for a short time τ, beginning at time t=0, and assume that daughters of labelled cells carry sufficient label to be detected as labelled themselves. The fraction of phase 4 cells which are labelled at time t is therefore just the fraction of phase 4 cells which were in phase 2 at any time during the initial pulse of labelling, and this is easily seen to be

$$(12) \qquad \mathrm{flm}(t) = \{\Psi_{3(4)}(t) - \Psi_{23(4)}(t-\tau)\} + \{\Psi_{34123(4)}(t) - \Psi_{234123(4)}(t-\tau)\} + \sum_{\nu=1}^{\infty}\{\Psi_{34;\nu;123(4)}(t) - \Psi_{234;\nu;123(4)}(t-\tau)\}$$

where the subscript 34;ν;123(4), for example, refers to the age, measured from the start of phase 3 of the (ν+1)st cycle prior to the present one, of a cell sampled from phase 4. Since flm(t) is relatively insensitive to assumptions concerning the variance of, or covariances with, phase 4, it is often sufficient to assume that phase 4 is independent of the other phases, in which case (12) simplifies to

$$(12a) \qquad \mathrm{flm}(t) = \{\Psi_{3(4)}(t) - \Psi_{23(4)}(t-\tau)\} + \sum_{\nu=1}^{\infty}\{\Psi_{3;\nu;(4)}(t) - \Psi_{23;\nu;(4)}(t-\tau)\}.$$

Computation of flm(t) is particularly simple in cases where the variability in phase 4 duration is negligible. Inspection of (12a), (11) and (6) shows that it is then only necessary to specify functional forms for the distributions h_3 (G_2 duration), h_{23} (S+G_2 duration) and h (cycle duration); fitting the curve to experimental data permits estimation of up to two parameters from each of these distributions.

Three examples of flm curves are shown in Fig. 1. The first rise of the curve is generated by $\Psi_{3(4)}(t)$, the first fall by $\Psi_{23(4)}(t-\tau)$, and so on. Roughly speaking, the position and slope of the first rise carry information about the mean and variance, respectively, of h_3, and similar statements hold for the other limbs of the curve. Estimation problems are discussed by Macdonald (1970).

Since only proliferating cells are counted in a pulse-labelling experiment, the parameters estimated by fitting a fraction labelled mitoses curve to the data refer to proliferating cells alone. However, in a situation where quiescent cells tend to become proliferative again the mean and variance of cycle duration will be inflated accordingly.

The experimenter may also measure the ratio of mitotic cells to all cells, called the mitotic index, at any time during the experiment, and the ratio of labelled cells to all cells, called the labelling index, just at the end of the pulse. Denoting these indices by MI and LI(τ), respectively, we find from (9), (2) and (5) that

$$\text{(13)} \qquad MI = (GF)\,(P_4) = \frac{GF}{A-1}\left\{\frac{1}{h_4^*(k)} - 1\right\}$$

and from (8), (2) and (5) that

$$\text{(14)} \qquad LI(\tau) = (GF)\,(P_2) = \frac{GF}{A-1}\,\frac{1}{h_4^*(k)}\left\{\frac{1}{h_{23}^*(k)} - \frac{e^{-k\tau}}{h_3^*(k)}\right\} .$$

By solving (13) and (14) while fitting the flm curve to data it is possible to estimate GF and one parameter of h_4 in addition to the parameters of h_3, h_{23} and h. Since slowly-cycling cells are accounted for by the variance of cycle duration, the growth fraction estimated in this manner includes "temporarily quiescent" cells with proliferating cells. The proportion of cells with short cycles would presumably be a somewhat smaller figure.

CELL PROLIFERATION IN A ROOT MERISTEM

Fig. 2 depicts a longitudinal section of a root meristem of *Zea mays*. Pulse-labelling experiments were carried out in various regions of the meristem, and Fig. 2 gives the values of growth fraction, mean cycle duration and mean duration of mitosis for these regions. These results may be compared to those in Barlow and Macdonald (1973), where the same data were used but the growth fraction was assumed to be 100% throughout the meristem.

It is clear from the analysis that the proliferative characteristics of the cells differ through the meristem. This casts some doubt on the validity of the model used, since the cells will change their positions relative to the quiescent centre as the root grows.

Possibly, the characteristics of cells within each of the three regions stele, cortex and epidermis depend only on distance from the quiescent centre, perhaps through some chemical gradient down the root. In this case it might be appropriate to set up a system of differential equations similar to those of Bartlett (1969), but including distance as an additional variable.

The initial cells of the cap are arranged in tiers and it appears that successive tiers are produced by a stem-cell population along the cap-quiescent centre boundary. It would not be difficult to set up a computer simulation of the cap

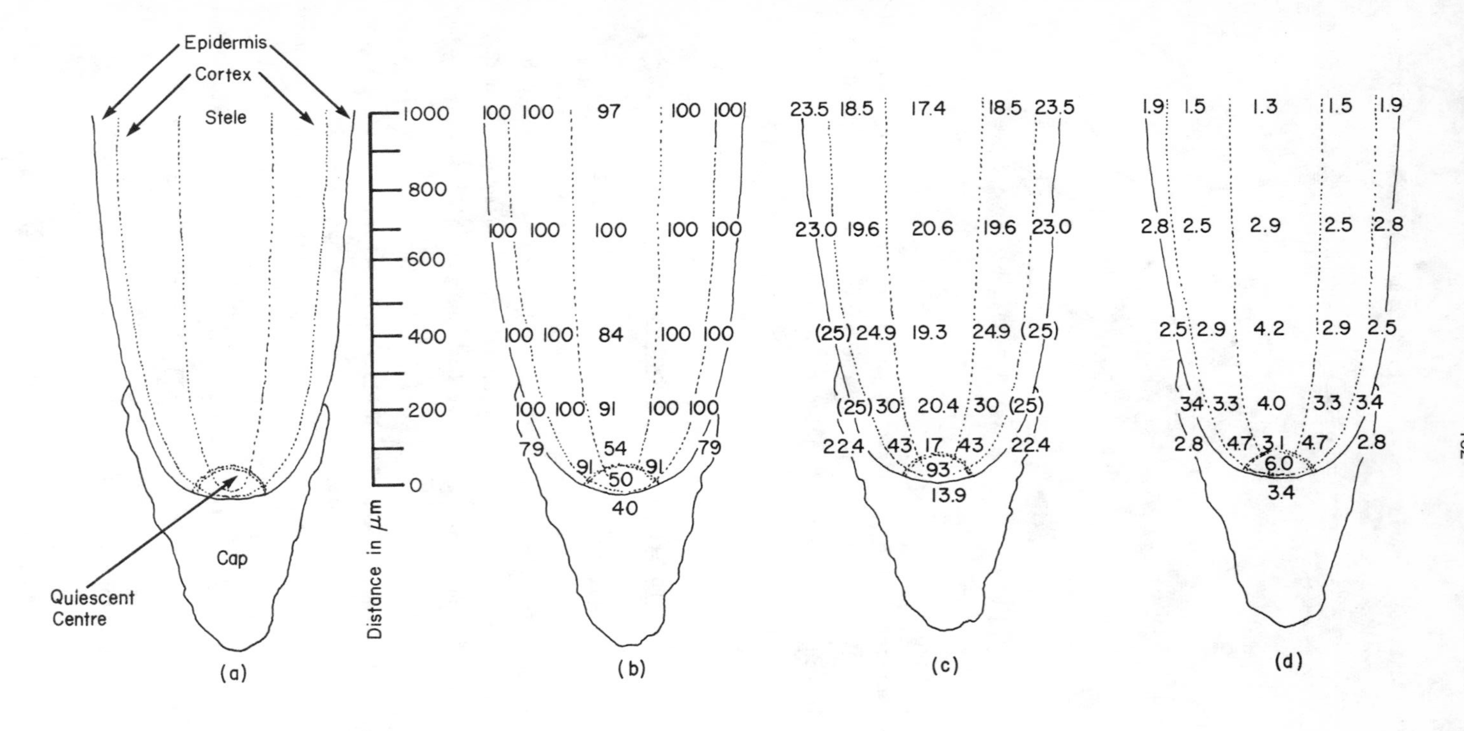

Figure 2. Proliferative characteristics of cells in various regions of the root meristem of Zea mays.
(a) The regions of the meristem; (b) growth fraction (%);
(c) Mean cycle duration (hr.); (d) Mean duration of mitosis (hr.).

region, allowing cells to change their properties as they move into successive tiers. Such a simulation could be used to predict the outcome of pulse-labelling and other similar experiments under assumptions quite different from those made previously.

I would like to thank Dr. P. W. Barlow of the Agricultural Research Council Unit of Developmental Botany, Cambridge, without whose persistence and enthusiasm this work would never get done. Figure 1 is reproduced from Biometrika with permission.

REFERENCES

BARLOW, P.W. and MACDONALD, P.D.M. (1973). An analysis of the mitotic cell cycle in the root meristem of *Zea mays*. *Proc. Roy. Soc.* B 183, 385-398.

BARTLETT, M.S. (1969). Distributions associated with cell populations. *Biometrika* 56, 391-400.

BROCKWELL, P.J. and KUO, W.H. (1972). Generalized asymptotic age distributions for a multiphase branching process. Statistical Laboratory Publication 38, Department of Statistics and Probability, Michigan State University.

CLOWES, F.A.L. (1971). The proportion of cells that divide in root meristems of *Zea mays* L. *Ann. Bot.* 35, 249-261.

HARRIS, T.E. (1963). *The Theory of Branching Processes*. Berlin: Springer - Verlag.

MACDONALD, P.D.M. (1970). Statistical inference from the fraction labelled mitoses curve. *Biometrika* 57, 489-503.

MACDONALD, P.D.M. (1973). On the statistics of cell proliferation. In *The Mathematical Theory of the Dynamics of Biological Populations* (Ed. R. Hiorns). London: Academic Press.

INSTRUCTIONAL LECTURE ON MATHEMATICAL TECHNIQUES

by

Z.A. MELZAK

Department of Mathematics
University of British Columbia
Vancouver, B.C.

ABSTRACT

This talk will be concerned with certain mathematical techniques as they enter into mathematical modelling of pattern-perception. Set-theoretic formulation of pattern-perception will be given and criticized, then techniques from topology, convexity, integral equations, and automata theory will be used to set up improved partial models. Some anatomical and physiological correlations will be discussed.

ANOMALOUS DIFFUSION THROUGH MEMBRANES

Grove C. Nooney
Lawrence Berkeley Laboratory
University of California, Berkeley.

ABSTRACT

Diffusive permeation through certain membranes has been reported to proceed anomalously faster toward increased surface concentrations. Such diffusion requires the existence of a fixed singular surface within the membrane where its diffusion coefficient becomes zero or infinite. Examples are provided. Part of this work appears in the Journal of the Chemical Society, Faraday Transactions II, 1973, vol. 69, at page 330.

The Role of Age Structure in the Dynamics of Interacting Populations

G. Oster
University of California
Berkeley, California

1. INTRODUCTION

The dynamics of populations may be treated on several levels. Most models of interacting populations have been at the overall population level, couched in terms of time invariant ordinary differential or difference equations. All populations, however, have an internal age structure which profoundly affects their dynamical response to external influences. Populations are intrinsically distributed parameter systems; to model them as finite dimensional systems is to assume a stable age distribution. Moreover, few populations live in a constant environment; both seasonal and circadian exogenous and endogenous periodicities militate against a pure and continuing Malthusian growth. Therefore, even granting a stable age distribution one is forced to abandon finite dimensional, autonomous equations as viable models. To relax either the assumption of stable age distribution or constant environments is to dramatically alter, even reverse, the qualitative and quantitative consequences of the model as they pertain to actual populations.

We have examined the dynamical effects of age structure and environmental periodicity on the behaviour of two interacting populations. Our central theme is that certain novel phenomena emerge as a natural consequence of viewing the populations as coupled, distributed parameter systems with periodic forcing.

In particular: (a) periodic solutions arise whose origin is quite different from that classically associated with population oscillations (May, 1972). (b) A hitherto unstudied phenomenon, that of "distributed resonances"--analogous to harmonic beats in linear systems--emerges as a candidate to explain long term periodicities and population outbreaks. (c) The "synchronization" of parasite-host observed experimentally by Hassel & Huffaker can be explained on the basis of our model. It is important to notice that these observations are fairly robust conclusions of a quite general model. The same qualitative behavior is observed over a wide range of parameter values.

Since this model is part of a larger biological control study in progress, we shall concern ourselves particularly with parasite-host systems. The results are easily generalizable to other types of population interactions.

2. POPULATION BALANCE MODEL

The equations describing the population density function have been derived in several contexts (Hulburt & Katz, 1964; Von Foerster, 1959; Sinko & Streiffer, 1967; Frederickson, Ramkrishna & Tsuchiya, 1967; Oldfield, 1966). We assume that the state of a particular individual in a population at time t is specified by the chronological age since birth, a, and a set of physiological parameters $\underline{\xi} = (\xi_1, \ldots, \xi_n)$. These may include size, mass, chemical compositions, or any other quantities having a bearing on the individual's rate of growth or reproduction. That is, an individual is represented by a point $(t,a,\underline{\xi}) \in \mathbb{R}^{n+2}$ and, as usual, we assume that knowledge of the state $(t,a,\underline{\xi})$ is sufficient to predict the trajectory of the individual according to the vector field:

$$\frac{da}{dt} = 1 \qquad (1) \qquad\qquad \frac{d\underline{\xi}}{dt} = \frac{d\underline{\xi}}{da} = \underline{g}(t,a,\underline{\xi}) \qquad (2)$$

where $\underline{g}$ is the growth rate for the qualities $\underline{\xi}$. A population of such individuals can be described by defining a set of density functions $n_k(t,a,\underline{\xi})$ on the state space, one for each population. The equations of motion for the population densities are obtained by applying a conservation law to each n_k:

$$\frac{\partial n_k}{\partial t} + \frac{\partial n_k}{\partial a} + \frac{\partial}{\partial \underline{\xi}} (\underline{g} n_k) = -\mu_k n_k \qquad (3)$$

where μ_k is a functional giving the death rate for population k. Initially, we shall restrict ourselves to two populations whose state is determined by time and chronological age alone.

The initial distributions for equations (3) are presumed known. The unique character of the demographic equations, vs similar appearing equations in other fields, is embodied in the death rate functional and in the boundary conditions which specify the sink and source terms for individuals. In particular, the boundary condition specifies the rate of entry of new individuals into the population at age zero as a functional of the adult breeding population. This is an integral feedback which typically takes the form $n_k(o,t) = \int m_k(t,a,\underline{n})\, n_k da$ where m_k is the "maternity function" which weights the contribution of the adult stock to the neonates. This feedback is illustrated schematically in fig. 1 for a single population.

The character of any population, insofar as its numerical abundance is concerned is completely determined by the birth and death rate functionals. With appropriate assumptions on these functionals, equations (3) reduce to most of the more common population models, eg. logistic, Gauss equations, Volterra-Lotka equations, Nicholson-Bailey, Leslie model, etc.

3. THE HOST-PARASITE SYSTEM

3.1 Consider a host population h(t,a) which is parasitized in one of its early instars (life stages) by a parasite population, p(t,a):

$$\frac{\partial h}{\partial t} + \frac{\partial h}{\partial a} = -\mu_1 h, \quad h(o,a) = f_1(a), \tag{4}$$

$$\frac{\partial p}{\partial t} + \frac{\partial p}{\partial a} = -\mu_2 p, \quad p(o,a) = f_2(a). \tag{5}$$

Let us examine the birth and death rate functionals for this system. The interactions between the two populations is represented schematically in fig. 2; parasitized hosts are numerically equal to the parasite neonates (assuming a host is parasitized only once). Therefore, the death rate of the host population will depend on (i) age, a; (ii) time in the season, t; (iii) the number of other hosts present per unit area (ie. density control), $\int h da$; (iv) the number of hosts eligible for parasitization, $\int \nu_1(a) h\, da$ & (v) the number of parasite adults eligible to attach the host, $\int \nu_2(a)pda$. Here ν_1, ν_2 are age-specific factors which select the eligible life stages and weight them appropriately. Therefore, μ_1 has the general form:

$$\begin{aligned} \mu_1 &= \mu_1 \, [t,a, \int hda, \int \nu_1 hda, \int \nu_2 pda] \\ &\triangleq \mu_1 \, [t,a,H,\tilde{H},\tilde{P}] \, . \end{aligned} \tag{6a}$$

The parasite death rate, while being age, time and density dependent, is not strongly dependent on the host density (this affects primarily the parasite birthrate)

$$\mu_2 = \mu_2 \, [t,a, \int \nu_3(a)pda]. \tag{6b}$$

Note that, while living within the body of the host, the parasite shares all external mortality causes of the host.

The birthrate functional for the host contains in addition to age, time and density dependence, a factor to account for the search capacity of the parasite. Since we are not treating spatial distribution effects explicitly in this model, we shall assume that the probability of parasitization is analogous to a Poisson process and employ a variant of the usual Nicholson-Bailey term (Hassel & Varley, 1967):

$$\text{probability of parasitization} \approx \tilde{H} \, (1-e^{-A\tilde{P}^{\gamma}}) \tag{7}$$

where A = parasite search area, $\tilde{H}$ & $\tilde{P}$ are the eligible life stages (eqn. 6a), and γ is a correction factor for nonrandom search capacity. The boundary conditions are of the form

$$h(o,t) = \int_o^\infty b_1(t,a)hda, \tag{8}$$

$$p(o,t) = \int_o^\infty b_2(t,a)pda \, . \tag{9}$$

Now, the birth and death rates of both parasite and host are seasonal. We introduce this in the form

$$b_1(t,a) = \sigma_1(t)b(a) \tag{10}$$

$$\mu_1(t,a,H,\tilde{H},\tilde{P}) = d_1(t)\mu_1(a,H,\tilde{H},\tilde{P}) \tag{11}$$

$$\mu_2(t,a,P) = d_2(t)\mu_2(a,P) \tag{12}$$

where σ_1, d_1, d_2 are periodic functions with the same frequency (but possibly different phases if the daily or seasonal cycle affects each species differently).

4. DISCUSSION OF EFFECTS

In many respects, the population equations (3) resemble models for other distributed parameter systems encountered in engineering, eg. crystal growth, convective heat exchange, etc. Therefore, we can expect at least as rich a universe of dynamical behavior from them, as well as some unique effects due to the integral feedback in the boundary conditions. We have concerned ourselves with the following phenomena:

(a) Because the seasonal forcing enters the birth and death terms in an age-specific fashion, waves will be excited which will propagate through the population profile. Although the nonlinearities will produce frequency multiplications, it is clear that a periodic signal which is an approximate multiple of a generation time will produce a resonance as the maxima from the previous period is reinforced each cycle. This phenomena of "distributed resonance" has been observed in other distributed parameter systems.

(b) In an analogous fashion, due to the age-specific nature of the population interactions, travelling waves will be excited in the host population which will, in turn, feed back and excite similar waves in the parasite population profile (see fig. 2). On an overall population scale, these will manifest themselves as an endogenous limit cycle oscillation, albeit of an entirely different origin than considered heretofore, being the result of the distributed nature of the system.

(c) These exogenous and endogenous periodic effects each produce travelling waves through the population profile. These waves will interact to produce maxima and minima in the populations whose spacing may be much longer than either a seasonal cycle or a generation time. That is, a phenomenon analogous to "beats" in harmonic systems is to be expected which will produce population outbreaks and crashes as the exogenously and endogenously excited waves reinforce and annihilate each other. We suggest that this last effect may account for certain long term periodicities in natural populations.

A third source of periodic excitation is in the differential effects of temperature and other abiotic factors on maturation rates. If, instead of chronological age, we employ a more meaningful physiological measure such as size, mass or chemical concentrations, then equations (4), (5) take the form

$$\frac{\partial n}{\partial t} + \frac{\partial}{\partial \xi}(gn) = -\mu n$$

where $g = g[T(t),\xi]$. Since growth rate is generally age specific, periodic temperature, T, variations can excite travelling waves in the population profile which can interact with those generated by the birth and death terms.

It is apparent from this discussion that the crucial time scales involved are the generation times of the species vs. the forcing frequency. Therefore, it is important to distinguish 3 types of populations with regard to long term periodic behaviour: (1) many generations per season, (2) approximately one generation per season, and (3) many seasons per generation.

The model possesses two interesting "bifurcation" phenomena when certain parameters are varied.

(i) As the "breeding window," γ, (c.f. fig. 1) is widened, at the point where $\gamma \sim \alpha$ = age at first reproduction, the root distribution of the system characteristic equation suddenly changes quite dramatically, resulting in a rapid change in the population growth rates. This provides a mechanism for the triggering of insect outbreaks by a sequence of favourable weather conditions which prolongs the reproductively active period.

(ii) As the interaction strength between host and parasite is varied, the entire system bifurcates to a periodic solution characterized by age profiles resembling "travelling waves": $n(a,t) = n(a,t+\tau)$. That is, the populations synchronize themselves, condensing their age structure into almost discrete generations. This effect has been observed experimentally and has important consequences for the timing of pest control measures. Once this synchronization has occurred, the system equations can be approximated by a coupled set of difference equations of the Nicholson-Bailey form. If the synchronization phenomenon is not rare in the insect world, this could account for the somewhat disconcerting success difference equation models have enjoyed even for ostensibly continuously breeding organisms.

The mathematical analysis of the model described by equations (4)→(12) is presented in Oster & Takahashi (1974) & Auslander, Oster & Huffaker (1974).

REFERENCES

D. Auslander, G. Oster, C. Huffaker (1974) J. Frank Inst. (to appear).

A.G. Frederickson, D. Ramkrishna, H. Tsuchiya (1967) Math. Biosci. 1, 327-74.

M. Hassell, G. Varley (1969) Nature, 1133-37.

M. Hassell, C. Huffaker (1969) Res. Pop. Ecol. 11, 186-210.

H. Hulburt, S. Katz (1964) Chem. Eng. Sci. 19, 555.

R. May (1972) Science, 177, 900-902.

D. Oldfield (1966) Bull. Math. Biosci. 28, 545-54.

G. Oster, Y. Takahashi (1974) Ecol. (to appear).

J. Sinko, W. Streifer (1967) Ecol. 48, 910-918.

H. Von Foerster (1959) in: The Kinetics of Cellular Proliferation, pp. 382-407, F. Stehlman, Jr., ed., New York: Grune & Stratton.

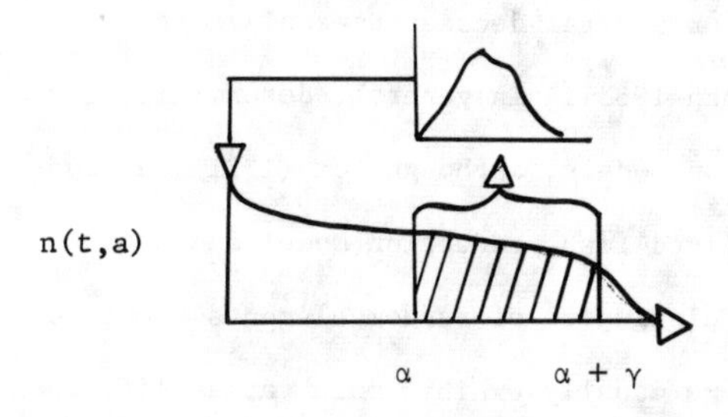

FIG. 1 Population profile & Birth feedback

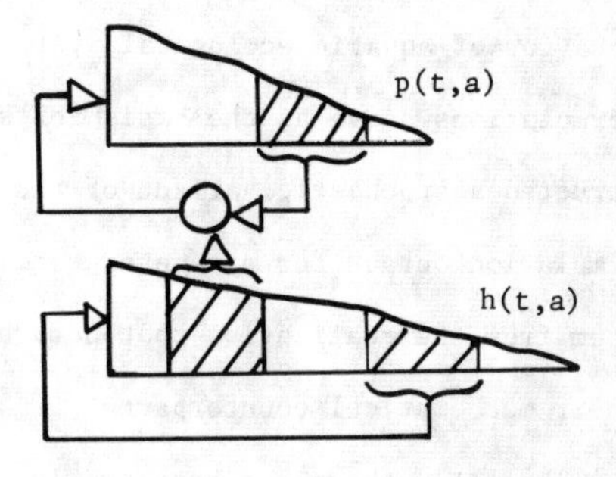

FIG. 2 Host-Parasite System

SOME CONSEQUENCES OF STOCHASTICIZING AN ECOLOGICAL SYSTEM MODEL

by

Richard A. Parker
Departments of Zoology and Computer Science
Washington State University
Pullman, WA 99163 U.S.A.

Differential equations have been used for several decades to analyze the behavior of aquatic ecological systems (Patten 1968). In general, deterministic formulations serve as the basis for simulation models, although Fox (1971) has constructed a stochastic variant of the generalized fish production model and used simulation output for parameter estimation. Inclusion of random elements seems to stem from the realization that natural systems usually exhibit more variability than their mathematical counterparts.

Recently there has been growing public attention focused on the effects of nutrient enrichment and industrial pollution (e.g. Lake Erie). Predicting these effects, together with the course of recovery following effluent treatment or diversion, has become the primary objective of many simulation models. Parker (1973) has devoted considerable effort to modeling the relationships among three nutrients (phosphate, nitrate, ammonium), two groups of algae, temperature, light intensity, and the zooplankton (copepods and cladocerans) in a large fjord-like lake. Results to date point to the stabilizing influence of a vast hypolimnetic nutrient pool, as well as to inadequate knowledge of the lake's hydrodynamic characteristics and nutrient cycling. Phytoplankton data collected thus far appear to be relatively "noisey", yet it is not obvious as to how much variation is real and how much could be removed by increased sampling intensity. Toward resolving this uncertainty, two stochastic versions of the 1973 deterministic model were developed and their performance will be discussed here.

This project has been financed in part with Federal funds from the United States Environmental Protection Agency under grant number R-800430.

THE MODELS

Each of the four biological components in the original model (I) was assigned specific growth and mortality functions (typically nonlinear), the difference being multiplied by the current density to describe total population change. These differential equations also included a transport term dependent on the massive seasonal influx of river water (flushing time approximately two years). Changes in the three nutrient concentrations were coupled to algal growth, zooplankton grazing, horizontal transport and vertical diffusivity. Field observations were made at five stations, one (STA 1) located at the river mouth and four (STA 2-5) spaced longitudinally along the lake. Multiple samples from three depths (1, 5, 10 m) were collected and mean values used to verify the model applied to an upper 10-m "mixed" layer. The differential equation system was treated as ordinary for each location by using observed values at the preceding station for estimating transport. Further details can be found in Parker (1973).

Each growth and mortality term in Model I contains a proportionality constant that serves to scale the output so that it is consistent in magnitude with actual observations. In Model II, 24 of these parameters were treated as random variables having the same means as those used in Model I, but subject to the possibility of up to plus or minus 20 percent variation. Modifications to all of these "constants" were made at the beginning of every interval in the numerical solution of the seven differential equations, utilizing pseudo-random numbers $\underline{R}$ from a uniform distribution.

Model III was developed to test the consequences of treating population growth and mortality as "birth and death" transitions in the sense of Bartlett (1960). This approach is, of course, a gross oversimplification since it is computationally impractical to deal with each organism as a separate entity, and migration (inward, outward) is allowed to occur only via deterministic transport terms between extremely large, adjoining segments of the lake. Transitions permitted for the two algal groups A_1 and A_2 and for the two groups of zooplankton C_1 and C_2 are summarized in TABLE 1. The $\underline{p's}$ and $\underline{n's}$ represent the nutrient content of total biomass--p_1 micromoles of phosphorus (N_1) per unit of phytoplankton and p_2 micromoles per unit of zooplankton; likewise n_1 and n_2 for nitrogen content. Growth G of each alga was made dependent on

TABLE 1. Birth and Death Transitions

A_1	A_2	C_1	C_2	N_1	N_2	N_3	$P/\Delta t$
+1	0	0	0	$-p_1$	$-n_1$	0	$A_1G_{A_1}N_2/(N_2+N_3)$
+1	0	0	0	$-p_1$	0	$-n_1$	$A_1G_{A_1}N_3/(N_2+N_3)$
0	+1	0	0	$-p_1$	$-n_1$	0	$A_2G_{A_2}N_2/(N_2+N_3)$
0	+1	0	0	$-p_1$	0	$-n_1$	$A_2G_{A_2}N_3/(N_2+N_3)$
-1	0	0	0	0	0	0	$A_1M_{1A_1}$
-1	0	0	0	p_1	0	n_1	$A_1TBc_{11}C_1(1-b_1)$
-1	0	+1	0	p_1-p_2	0	n_1-n_2	$A_1TBc_{11}C_1b_1$
-1	0	0	0	p_1	0	n_1	$A_1TBc_{21}C_2(1-b_2)$
-1	0	0	+1	p_1-p_2	0	n_1-n_2	$A_1TBc_{21}C_2b_2$
0	-1	0	0	0	0	0	$A_2M_{1A_1}$
0	-1	0	0	p_1	0	n_1	$A_2TBc_{12}C_1(1-b_1)$
0	-1	+1	0	p_1-p_2	0	n_1-n_2	$A_2TBc_{12}C_1b_1$
0	-1	0	0	p_1	0	n_1	$A_2TBc_{22}C_2(1-b_2)$
0	-1	0	+1	p_1-p_2	0	n_1-n_2	$A_2TBc_{22}C_2b_2$
0	0	-1	0	0	0	0	$C_1M_{1C_1}$
0	0	-1	0	0	0	0	$C_1M_{2C_1}$
0	0	0	-1	0	0	0	$C_2M_{1C_2}$
0	0	0	-1	0	0	0	$C_2M_{2C_2}$

$\Delta t = -\log(R)/\lambda$ $\qquad$ $\lambda = \Sigma P/\Delta t$

total nitrogen and partitioned here into that accounted for by nitrate (N_2) and ammonium (N_3). T represents temperature, B is a grazing activity coefficient which decreases exponentially with algal density, c's are selective grazing coefficients, b's are fractions of phytoplankton eaten converted to zooplankton, and M is a mortality function (M_1 = natural, M_2 = predation). The unit of population chosen for simulation was 10 micrograms since smaller values did not alter performance pattern but did increase substantially the computing time required.

RESULTS AND DISCUSSION

Output from Models II and III for phosphate, two algal groups, and cladocerans is presented in FIGURES 1-3, respectively, for three years. That for nitrate, ammonium, and copepods has not been included since the other components satisfactorily illustrate patterns of general system behavior. Model II simulation results for stations 2-5 are shown in the lower half of each figure--results from Model III in the upper half. Comparison of Model I output (Parker 1973) with that from Model II dramatically demonstrates that randomization of proportionality constants for each computational interval had virtually no effect. It must be emphasized, however, that this conclusion should not be extended to include all parameters in the model without further study.

On the other hand, use of birth and death transitions (Model III) caused little difference in simulated phosphate values (FIGURE 1) but exceptionally abnormal behavior of the phytoplankton (FIGURE 2) and cladoceran populations (FIGURE 3). The algal group (2) which ordinarily dominates during the spring is permanently replaced by the late summer group (1) after the first year. Of further interest is the fact that the cladoceran population begins to wane even though its members feed preferentially on the now dominant summer algal forms. Rapid extinction occurs at station 2 since no lake plankton is assumed to enter the lake in river water. Results at stations 3-5 should therefore be viewed as more representative.

Although one is forced to conclude that Model III is not a satisfactory representation of the real system, it may be useful to look for signs of instability in the underlying model (I). May (1972), based on work by Kolmogorov on autonomous systems, suggests that those natural ecosystems which seem to exhibit a persistent

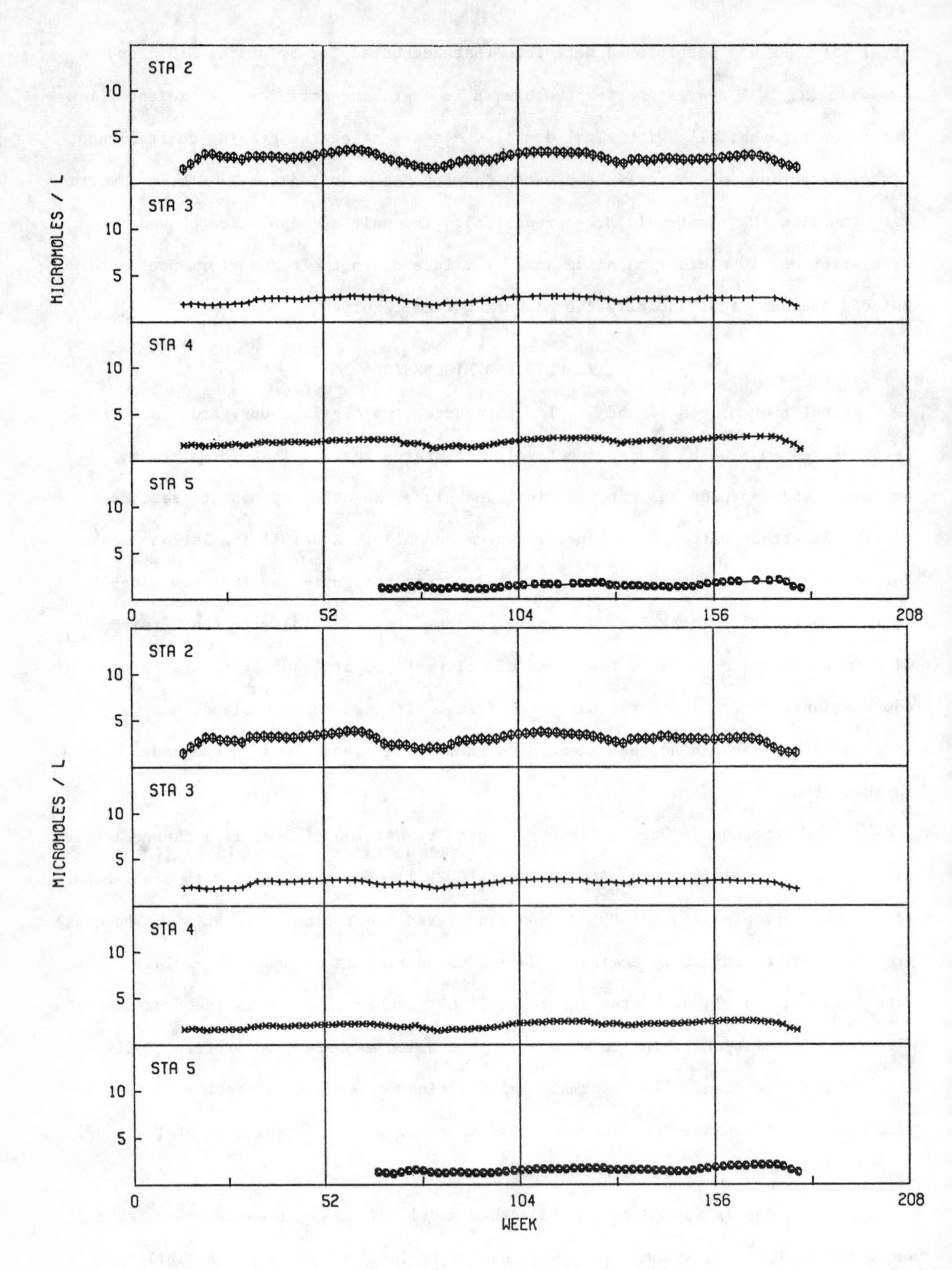

FIGURE 1. Simulated Phosphate Concentrations--Model II (lower) and Model III (upper)

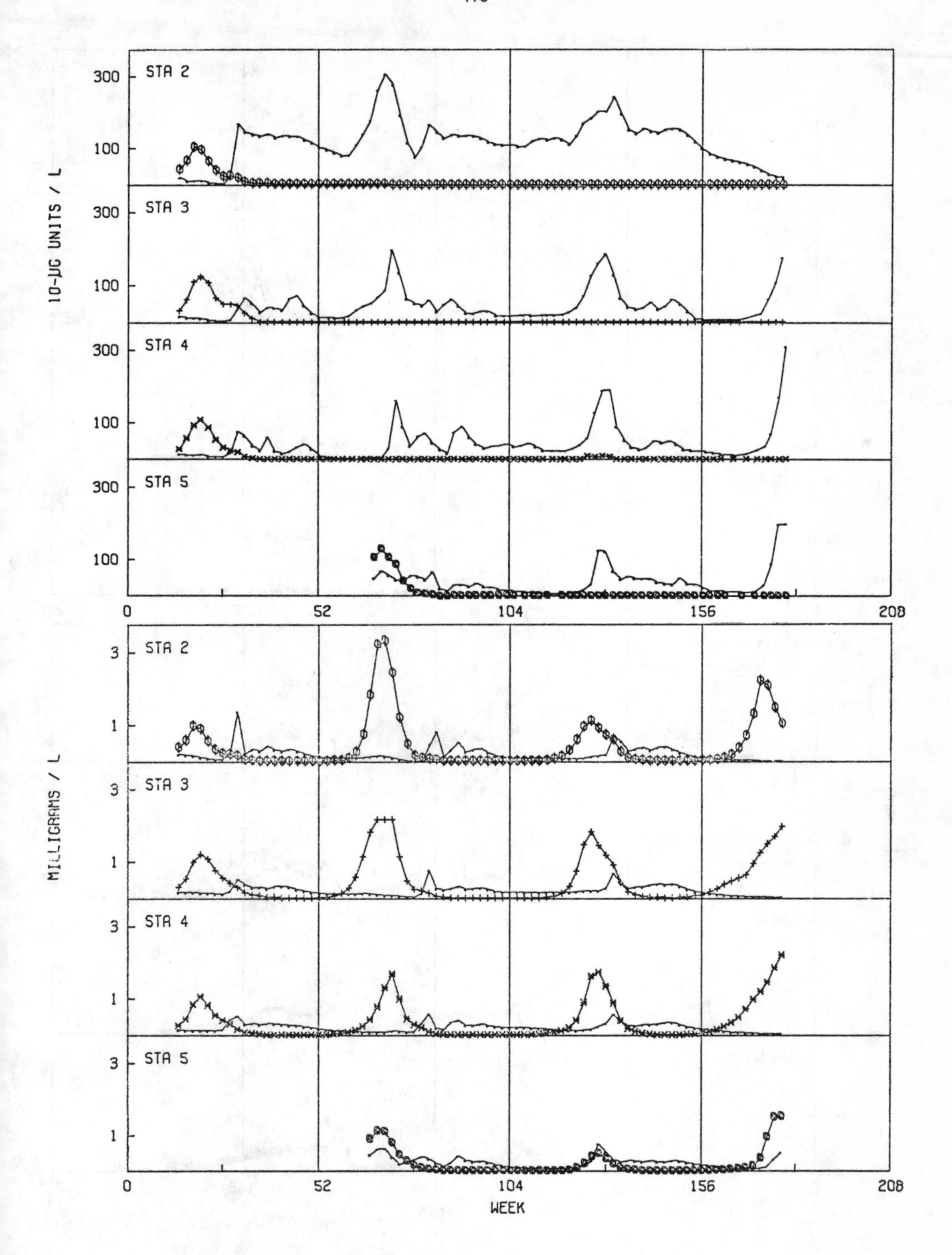

FIGURE 2. Simulated Phytoplankton Densities--Model II (lower) and Model III (upper)

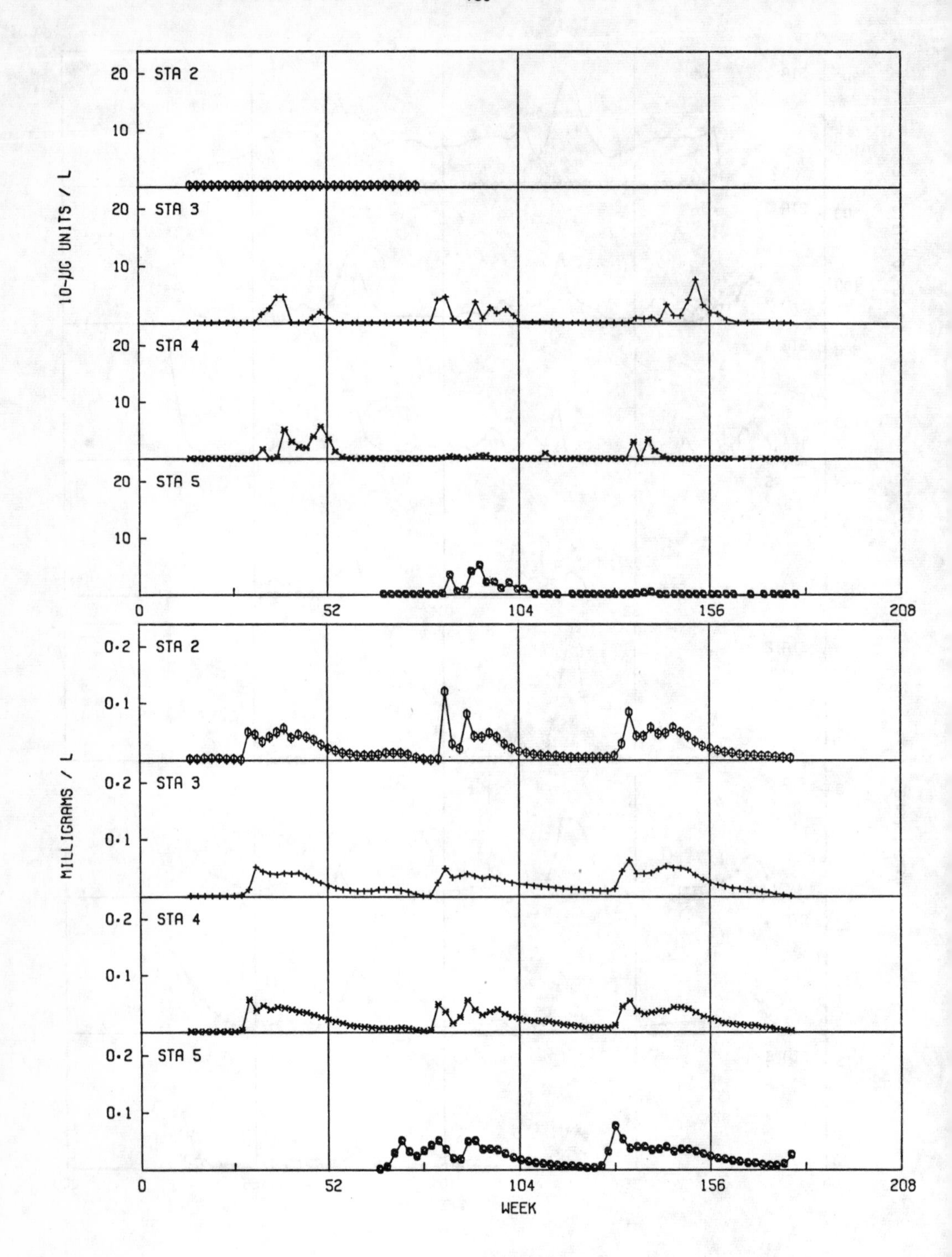

FIGURE 3. Simulated Cladoceran Densities--Model II (lower) and Model III (upper)

pattern of reasonably regular oscillations have in fact stable limit cycles. Three-year solutions based on Model I are stable in that they are bounded and there is no change in dimensionality. Nevertheless, if one periodically computes characteristic roots using variational equations, the number of negative real parts might give some measure of the tendency toward instability at various points during simulation. This has been done for station 2, and the results are given monthly in FIGURE 4. Note that the number of negative roots decreases markedly during periods of rapid algal growth, yet there is no good indication of why group 2 became extinct in Model III rather than group 1.

Unfortunately the equations employed here are nonautonomous in that growth and mortality rates are functions of light intensity and/or temperature; therefore, significant changes in these physical variables near a point in question could completely invalidate conclusions drawn from an analysis designed for autonomous systems. Perhaps Minorsky's review (1962) of Floquet Theory and Liapounov's Second Method is somewhat more relevant, since it considers the stability of linear systems with periodic coefficients. If so, the number of roots with modulus greater than 1 would be a better measure of potential instability. The difficulty now becomes one of identifying the proper characteristic equation. Failing this, the number of such roots calculated from the equation used in the autonomous case above could conceivably provide a measure of the intensity of activity at a point, rather than direction. These numbers (when non-zero) have been placed immediately above the number of negative roots in FIGURE 4. Interestingly, their occurrence is confined largely to late summer periods. One must bear in mind, however, that this phenomenon may only reflect the absence of higher order terms in the variational equations. None the less, further modification and refinement of the approach could conceivably aid in predicting major instabilities in natural ecosystems (e.g. massive summer algal mortality followed by oxygen depletion accompanying decomposition).

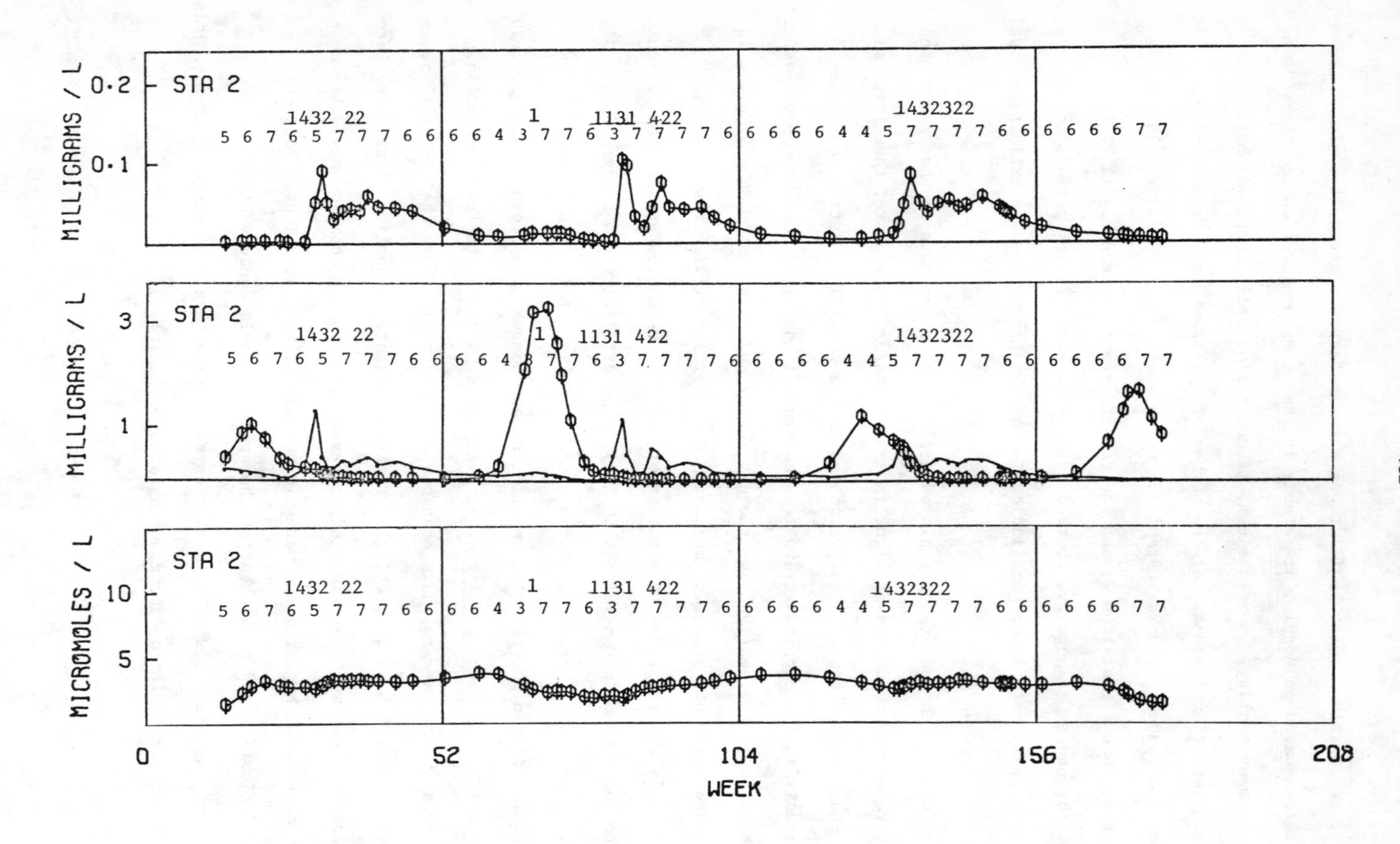

FIGURE 4. Simulated Values from Model I for Phosphate (lower), Phytoplankton (middle), and Cladocerans (upper)

REFERENCES

Bartlett, M. S. Stochastic Population Models in Ecology and Epidemiology. Methuen, London (1960)

Fox, W. W. Fish. Bull. 69(3), 569-580 (1971)

May, R. M. Sci. 177, 900-902 (1972)

Minorsky, N. M. Nonlinear Oscillations. D. van Nostrand, Princeton (1962)

Parker, R. A. Mathematical Theory of the Dynamics of Biological Populations. M. S. Bartlett and R. W. Hiorns (ed.) Academic Press, London (1973)

Patten, B. C. Int. Rev. Ges. Hydrobiol. 53(3), 357-408 (1968)

COMPETITION ON AN ENVIRONMENTAL GRADIENT

E. C. PIELOU

Biology Department, Dalhousie University,

Halifax, Nova Scotia

SUMMARY

A simple model is proposed to describe competition between two species on an environmental gradient. There are two versions of the model. One, which explains the gradual blending of adjacent zones sometimes found in zoned vegetation, postulates the existence of transition zones in which species can coexist in stable equilibrium. The other, which explains abrupt zonation, postulates the existence of transition zones in which the equilibrium is unstable.

A method of simulating the operation of the model is given. It allows for different intensities of within- and between-species competition in the different age classes of the competitors.

Some numerical results obtained by simulation are described. They suggest some predictions concerning the age distributions to be expected during the successional stages occurring while an area with a strong environmental gradient is being colonized. Temporary species-populations (destined to be the victims of competitive exclusion) tended to have disproportionately large numbers of young members for a longer time than winning populations, whose age distributions approached stability comparatively fast.

INTRODUCTION

There has been a recent resurgence of interest among population ecologists in "simple" models purporting to describe the dynamics of many-species populations (see, for example, Keyfitz 1968, Levins 1968, MacArthur 1970, Levin 1970, Vandermeer 1970, May 1973, Strobeck 1973). Much attention has been given to equations of the form

$$\dot{N}_i(t) = r_i N_i(t) - \frac{r_i N_i(t)}{K_i} \sum_{j=1}^{k} a_{ij} N_j(t) \qquad (1)$$

for $i = 1, 2, \ldots, k$, which are intended to describe the growth, in an environment with limited resources, of populations of k competing species. Here $N_i(t)$ is the size of the ith species-population at time t ; r_i is the intrinsic rate of natural increase of the ith species; K_i is the maximum sustainable population size of the ith species in the absence of other species. And a_{ij} is a competition coefficient; it measures the effect of the jth species' presence on the ith species' growth relative to the effect of the ith species on its own growth; (thus $a_{ii} = 1$).

Research into the properties of this model has been motivated in large part by an interest in its stability properties (e.g. Levin 1970, May 1973) and by its bearing on enquiries into the number of species that can coexist in a limited environment (e.g. MacArthur 1970, Strobeck 1973). In most of this work the simple assumptions underlying the model have not been called in question. It has been tacitly assumed that the qualitative conclusions arrived at may safely be accepted even though the model does not (but for an exception noted below) take account of the following four "complications": stochastic occurrences; delayed responses; the age structures of the competing populations; and the fact that in nature population growth is rarely continuous [but May (1973) does consider discretized as well as continuous growth rates and compares their consequences].

It therefore seems worth while to consider whether and in what circumstances natural populations whose behavior can be closely described by this model, or by a somewhat more realistic version of it, do in fact exist; and how they may be recognized. The model rests on two other simple assumptions (besides the four mentioned above) which seem at first sight to ensure the failure of any search for realizations of it in natural, as distinct from laboratory, communities. These are: that the competing organisms cannot relieve the effects of overcrowding by emigration; and that habitat conditions are constant throughout the area occupied by the populations. There is no objection to the first of these assumptions if the organisms concerned are sessile. As to the second, there is no need to make it.

For suppose we deliberately search for places with a strong, unidirectional, environmental gradient; then the habitat heterogeneity that such a gradient entails becomes an asset rather than a liability as will be shown. The purpose of this paper is to explore the effects of environmental gradients on competition between plant species; and to describe computer simulations of the process whose results suggest what observations should be made in the field to decide whether the model is to be accepted or rejected.

COMPETITION ON AN ENVIRONMENTAL GRADIENT

For simplicity we consider competition between only two species, called M and N.

Putting $r_M/K_M = s_M$; $a_{MN}s_M = u_M$;

$r_N/K_N = s_N$; $a_{NM}s_N = u_N$,

equations (1) become

$$(1/M)\,dM/dt = r_M - s_M M - u_M N \quad \text{and} \quad (1/N)\,dN/dt = r_N - s_N N - u_N M .$$

The possible outcomes predicted by this competition model are well known. Species M will win and N become extinct if

$$r_M s_N > r_N u_M \quad \text{and} \quad r_M u_N > r_N s_M \ .$$

The opposite outcome is inevitable (assuming a deterministic model) if both these inequalities are reversed.

However if one of the inequalities is reversed but not the other, so-called equilibrium states are possible. Thus

$$r_M s_N > r_N u_M \quad \text{and} \quad r_M u_N < r_N s_M$$

are the conditions for stable equilibrium in which the two species can coexist permanently. These conditions are shown graphically in the inset figure in the lower right corner of Figure 1. The straight lines connect all points (M,N) at which $dM/dt = 0$ (solid line) and $dN/dt = 0$ (dashed line). The curved dotted lines are representative trajectories of mixed populations in the M-N phase space. Conversely

$$r_M s_N < r_N u_M \quad \text{and} \quad r_M u_N > r_N s_M$$

are the conditions for unstable equilibrium. The setup is shown graphically in the inset figure in Figure 2. As before, the solid and dashed straight lines show population compositions at which $dM/dt = 0$ and $dN/dt = 0$ respectively. Permanent coexistence is impossible; which species will win, and which become extinct, depends on the species composition of the combined population initially. Thus, depending on their starting points, the dotted trajectories approach the abscissa ("M wins") or the ordinate ("N wins").

This model is familiar to all ecologists. Any one realization of the process is a temporal sequence of events, presumed to take place in a homogeneous environment. Now let us envisage a heterogeneous environment, specifically an environmental gradient along which some overwhelmingly important habitat factor varies unidirectionally. Graphical representation of the setup then requires a three-dimensional

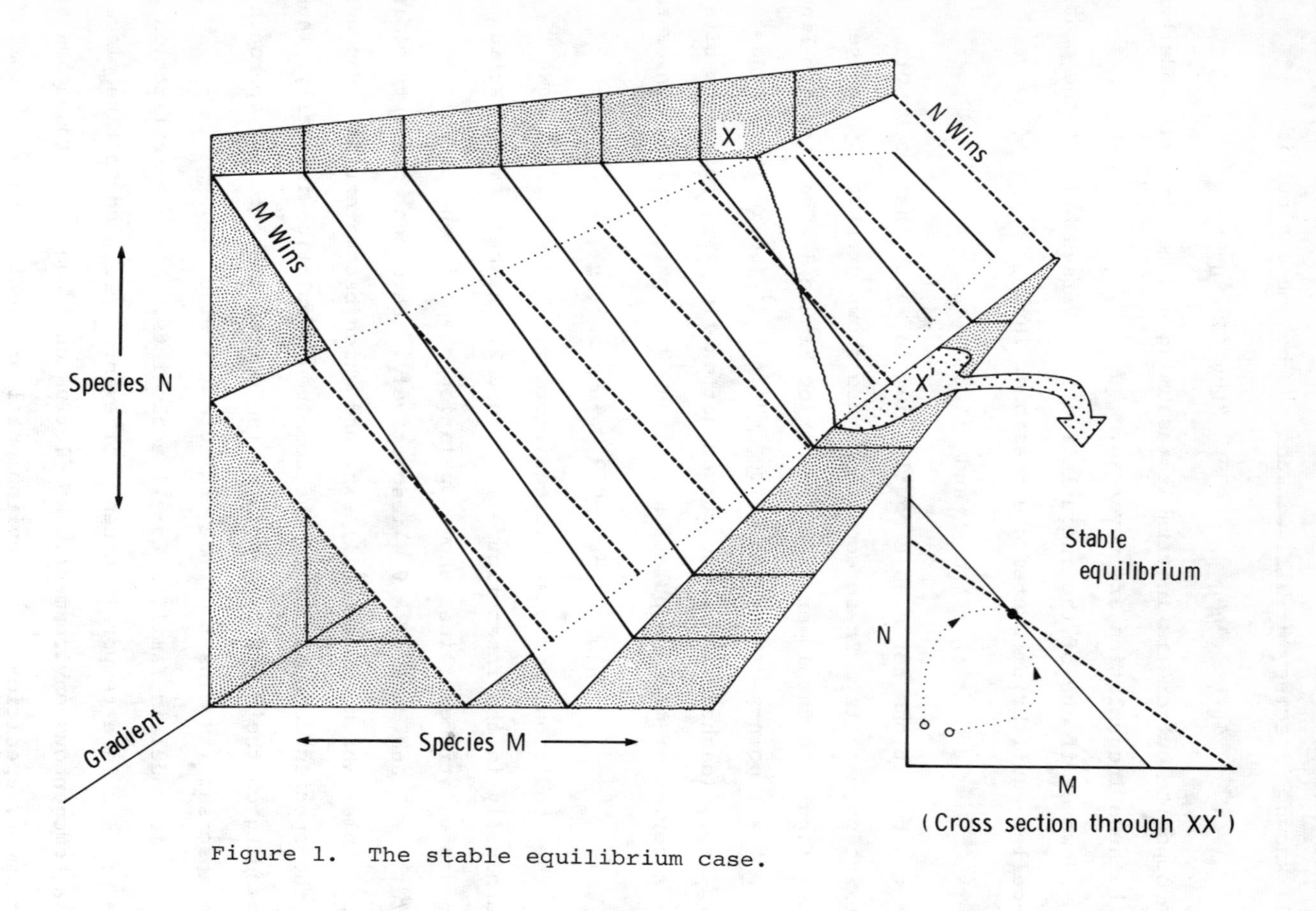

Figure 1. The stable equilibrium case.

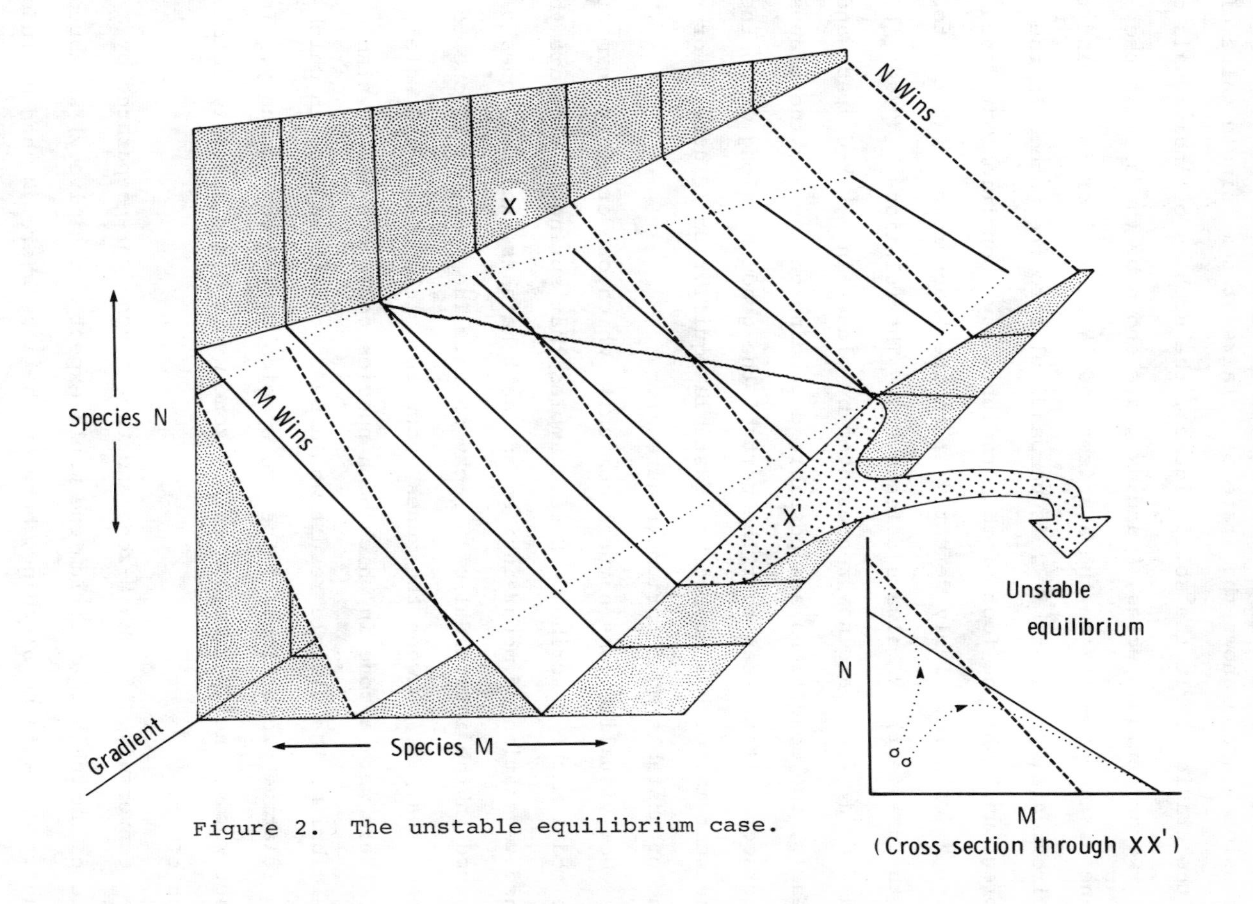

Figure 2. The unstable equilibrium case.

figure as shown in Figures 1 and 2. In each of these the inset two-dimensional graph shows the state of affairs at some single value of G, the gradient. In the solid figures, the numbers of individuals of the two competing species, M and N, are shown by the axes in the plane of the page (M on the abscissa and N on the ordinate) and the environmental gradient, G, is perpendicular to the page. In the "foreground" (at the foot of the gradient, say) conditions are such that M will invariably defeat N; in the "background" (at the top of the gradient) N always defeats M. The relations $dM/dt = 0$ and $dN/dt = 0$ are now represented by planes in M-N-G space and, clearly, two conditions are possible depending on the way the planes intersect. It is seen that the part of the gradient along which they intersect may either be a zone of stable equilibrium (Figure 1) or a zone of unstable equilibrium (Figure 2).

Now visualize some natural community which one or other form of the model might describe. Some examples that spring to mind are the zoned vegetation of mountainsides and salt marshes, and the zones of seaweed in the intertidal of a rocky shore. The model leads one to expect that if, between the zones occupied solely by M and solely by N, there were a zone in which both species could coexist in stable equilibrium, the visible result would be a transition zone in which the relative proportions of the two species varied continuously. In other words, the M-zone and the N-zone would blend smoothly into each other.

Conversely, if the M-zone and the N-zone were separated by a zone of unstable equilibrium one would expect (intuitively) an abrupt transition, with no blending, between the zone where M had excluded N and the zone where N had excluded M. It should be noticed, however, that the model does not lead to this last conclusion without some elaboration since, as it stands, it does no more than predict the ultimate fate, at any point along the gradient, of a given starting population. For there to be an abrupt transition between the M-zone and the N-zone with no blending, it must be assumed that on M's side of the transition line (which bisects the zone of unstable equilibrium) the mixture of disseminules that settle and germinate always contains M's and N's in such proportions as to ensure a trajectory leading to M's victory over N and conversely on N's side of the transition line. One would expect this to happen in any case, especially if the unstable zone were narrow and the disseminules heavy.

There is another possible explanation for the existence of an abrupt discontinuity between zones. There would be one if the $dM/dt = 0$ and $dN/dt = 0$ lines were parallel in all planes at right angles to the G-axis. The line of intersection of the two planes would then be of the form $aM + bN + c = 0$, and the length of its projection on the G-axis (which is the length of the zone of equilibrium) would therefore be zero. In this limiting case the intersection of the planes is the line where $dM/dt = 0$ and $dN/dt = 0$ are coincident with each other.

The latter possibility is by no means inherently unlikely if the competing species are closely related. Thus the geometrically precise zonation one finds in the seaweeds of steep rocky shores in sheltered locations may provide examples. Consider two closely related species, of the rockweed genus Fucus for example, that grow in contiguous zones. The taxonomic closeness of the species and the contiguity of their zones both suggest that their environmental requirements are extremely similar. Between-species interactions would probably differ only slightly from within-species interactions, and along the line where the zones come in contact one would expect to find that the limiting density for both species was the same, i.e. that $K_M = K_N = K$, say. This is also a necessary condition for there to be no discontinuity in biomass per unit area at the boundary between the zones, which seems a reasonable assumption. Thus if $K_M = K_N = K$ and $s = u$ for both species, the intersection of the $dM/dt = 0$ and $dN/dt = 0$ planes is the line $M + N = K$, and the equilibrium zone has vanished. This special case will not be considered further in what follows.

SIMULATION OF THE MODEL

To test whether the model does indeed explain the growth processes leading to zonation of vegetation it is clearly necessary to derive some other consequences of the model (other than zonation, that is) and discover, by field observation, whether these other consequences are in fact realized. In the simple form so far considered, however, the model makes no other predictions and in order to make some it is necessary to elaborate the model somewhat. This also has the advantage of making it more realistic. The predictions of the improved model can then be found by computer simulation.

To carry out this program, I modified the simple model in two ways: (i) The age distributions of the two competing species were taken into account; it was assumed that the effects on survival rates of within- and between-species crowding differed in different age classes. For simplicity, fertility rates were assumed to be unaffected by crowding. (ii) Time was treated as discrete rather than continuous. The model's predictions could then be generated by use of a projection matrix (Leslie 1945, Keyfitz 1968) which, at any given point on the environmental gradient was as follows:

$$Q_t = \left[\begin{array}{c|c} A & 0 \\ \hline 0 & B \end{array}\right]$$

where 0 is the $(k+1)\times(k+1)$ zero matrix,

$$A = \begin{bmatrix} F_0 & F_1 & \cdots & F_{k-1} & F_k \\ P_0(M,N) & 0 & \cdots & 0 & 0 \\ 0 & P_1(M,N) & \cdots & 0 & 0 \\ \cdots & \cdots & \cdots & \cdots & \cdots \\ 0 & 0 & \cdots & P_{k-1}(M,N) & 0 \end{bmatrix}$$

and

$$B = \begin{bmatrix} f_0 & f_1 & \cdots & f_{k-1} & f_k \\ p_0(M,N) & 0 & \cdots & 0 & 0 \\ 0 & p_1(M,N) & \cdots & 0 & 0 \\ \cdots & \cdots & \cdots & \cdots & \cdots \\ 0 & 0 & \cdots & p_{k-1}(M,N) & 0 \end{bmatrix}.$$

(It should be emphasized that the values of some of the elements in the matrices are to be made to vary along the environmental gradient. However, until further notice, we shall be considering the process at some one arbitrary point on the gradient.)

It is convenient to take the unit of time as one year. Both species are assumed to live for $k+1$ years.

The elements of Q_t are defined as follows:

F_j denotes the number of offspring born to a j-year old M-individual that survive to the end of the year in which they were born. They become established close to the parents.

$P_j(M,N)$ denotes the proportion of j-year old M-individuals that survive to reach the age $j+1$; this proportion is a function of M_t and N_t , the current sizes (at time t) of the competing populations.

The corresponding quantities for N-individuals are f_j and $p_j(M,N)$.

The functions $P_j(M,N)$ and $p_j(M,N)$ are defined thus:

$$P_j(M,N) = \Theta_j/(1 + \alpha_{Mj}M_t + \gamma_{Mj}N_t)$$

$$p_j(M,N) = \theta_j/(1 + \alpha_{Nj}N_t + \gamma_{Nj}M_t).$$

The α's and γ's measure the decrease in probability of survival caused by within-species and between-species competition respectively. The first subscript (M or N) denotes the species, and the second ($j = 0,1,\ldots$ or $k-1$) the age, of the affected individuals.

The age structure of a mixed population of M's and N's at time t can be represented by a column vector, V_t say, where

$$V_t' = (m_{0t}, m_{1t}, \ldots, m_{kt}, n_{0t}, n_{1t}, \ldots, n_{kt}).$$

Here m_{xt} is the number of x-year old individuals of species M present at time t, and correspondingly for n_{xt}. The total sizes of the two populations at time t are

$$M_t = \sum_x m_{xt} \quad \text{and} \quad N_t = \sum_x n_{xt} .$$

The sizes and age distributions of the two competing populations, at a sequence of times, can then be modeled by repeated application of the equation

$$V_{t+1} = Q_t V_t .$$

Now let us envisage a simple concrete example in order to generate some numerical results for inspection. The following assumptions will be made:

(i) That the two species have the same fertilities.

(ii) That their life span is two years.

(iii) That every young plant would survive into its second year if it were not for the effects of the competitors. This amounts to assuming that $\Theta_0 = \theta_0 = 1$.

Making these assumptions, Q_t becomes

$$
Q_t = \begin{bmatrix} F_0 & F_1 & 0 & 0 \\ P(M,N) & 0 & 0 & 0 \\ 0 & 0 & F_0 & F_1 \\ 0 & 0 & p(M,N) & 0 \end{bmatrix}
$$

$$
\begin{aligned}
\text{where} \quad & P(M,N) = 1/(1 + \alpha_M M_t + \gamma_M N_t) \\
\text{and} \quad & p(M,N) = 1/(1 + \alpha_N N_t + \gamma_N M_t) \quad .
\end{aligned} \tag{2}
$$

Next, let us choose numerical values for the F's. Absolute values are not required if we imagine the quantities of each species to be measured by, say, amounts per unit area and leave the areal unit unspecified. We must, however, decide on their relative values and shall suppose that fertility in the second year is twice that in the first. Hence we put $F_0 = 0.5$ and $F_1 = 1.0$.

To complete the specification of Q_t we now require the α's and γ's. As will be shown, if the population sizes of the M's and N's at equilibrium are given, only the ratios γ_M/α_M and γ_N/α_N can be chosen at will. To provide an unstable model we shall put

$$
\gamma_M/\alpha_M = 2.5 \quad \text{and} \quad \gamma_N/\alpha_N = 2.0 \quad . \tag{3}
$$

The absolute values of the α's and γ's are then determined by the fact that a combined population whose growth is governed by Q_t must ultimately become stationary (i.e. stable in age structure and constant in total size); or equivalently, that the dominant latent root of $\lim_{t\to\infty} Q_t = Q$, say, be unity.

Let us put $\lim_{t\to\infty} P(M,N) = \Phi$ and $\lim_{t\to\infty} p(M,N) = \phi$.

Then the characteristic equation $|Q - \lambda I| = 0$ becomes

$$
(\lambda^2 - F_0\lambda - F_1\Phi)(\lambda^2 - F_0\lambda - F_1\phi) = 0 \ .
$$

For the identical pair of dominant latent roots to be unity, it is necessary that $\Phi = \phi = 0.5$. We now have numerical values for all the elements of $\mathbf{Q}$.

The stable age distributions of the two species are given by the dominant latent vector of $\mathbf{Q}$ which is

$$\lim_{t\to\infty} \mathbf{V}_t' = \lim_{t\to\infty} (m_{0t} \quad m_{1t} \quad n_{0t} \quad n_{1t}) = (2 \quad 1 \quad 2 \quad 1) \quad .$$

Now return to the problem of assigning numerical values to the α's and γ's. Recall that we are still envisaging events at some arbitrary point on the environmental gradient. Assume that at this point a mixed population of the two species will be in equilibrium if $M = M_e$ and $N = N_e$. (The equilibrium could be stable or unstable). This is equivalent to assuming that (in the G = const. plane we are considering) the lines $dM/dt = 0$ and $dN/dt = 0$ intersect at (M_e , N_e). Then the α's and γ's are obtained by solving the equations:

$$1/\Phi = 1 + \alpha_M M_e + \gamma_M N_e = 2 \quad \text{and} \quad 1/\phi = 1 + \alpha_N N_e + \gamma_N M_e = 2 \; .$$

Hence, on substituting from (3),

$$\alpha_M = 1/(M_e + 2.5N_e) \quad \text{and} \quad \alpha_N = 1/(N_e + 2M_e) \; . \qquad (4)$$

We are now in a position to simulate the competition process at the chosen point on the environmental gradient. This is done by repeatedly premultiplying some chosen initial vector, say

$$\mathbf{V}_0' = (m_{00} \quad m_{10} \quad n_{00} \quad n_{10}),$$

by the projection matrix

$$\mathbf{Q}_t = \begin{bmatrix} 0.5 & 1.0 & 0 & 0 \\ P(M,N) & 0 & 0 & 0 \\ 0 & 0 & 0.5 & 1.0 \\ 0 & 0 & p(M,N) & 0 \end{bmatrix} \; .$$

The elements $P(M,N)$ and $p(M,N)$, whose values are obtained from (2),

(3) and (4), are recalculated at every step to allow for the current population sizes M_t and N_t .

In this way one may predict the progress of competition at any point on the gradient for which the values of M_e and N_e are specified. Thus final specification of the model consists in writing down the equations of the intersecting planes for which $dM/dt = 0$ and $dN/dt = 0$. To obtain numerical results the planes were defined as follows:

$$2M + 5N + 2G = 250 \qquad \text{(a)}$$

$$10M + 5N - 2G = 200 \quad . \qquad \text{(b)}$$

They intersect along the line

$$6(M - 6.25) = -5(N - 37.5) = 3(G - 25)$$

which cuts the $M = 0$ plane at $G = 12.5$
and the $N = 0$ plane at $G = 87.5$.

Therefore there is a transition zone over the range $12.5 < G < 87.5$. If the planes are defined so that (a) is the plane $dM/dt = 0$ and (b) is the plane $dN/dt = 0$, then the transition zone is one of unstable equilibrium. Everywhere in the zone $\alpha_M\alpha_N < \gamma_M\gamma_N$. Also, for $G < 12.5$ M always wins and for $G > 87.5$ M always loses.

Conversely, if (a) is $dN/dt = 0$ and (b) is $dM/dt = 0$, there is stable equilibrium in the transition zone. The values of the α's and γ's are interchanged so that $\alpha_M\alpha_N > \gamma_M\gamma_N$. For $G < 12.5$, M always loses and for $G > 87.5$ M always wins.

Table 1 (on the following page) summarizes the conditions used for the simulations. They were performed at points on the gradient given by $G = 25$, 50 and 75 respectively. For each of these chosen values of G , M_e and N_e were found by solving the pair of equations (a) and (b). Then the α's and γ's were obtainable from (3)

TABLE 1

SPECIFICATIONS OF THE MODELS USED TO SIMULATE TWO-SPECIES COMPETITION ALONG A GRADIENT

Parameters (For unstable model)	Gradient Level			Parameters (For stable model)
	$G = 25$	$G = 50$	$G = 75$	
α_M	0.010	$0.01\dot{3}$	0.020	γ_M
α_N	0.020	$0.01\dot{6}$	0.0143	γ_N
γ_M	0.025	$0.03\dot{3}$	0.050	α_M
γ_N	0.040	$0.03\dot{3}$	0.0286	α_N
M_e	6.25	18.75	31.25	
N_e	37.50	22.50	7.50	

and (4).

The object of performing simulations is to search for clear diagnostic differences between the course of events predicted by the model in a zone of stable equilibrium and that in a zone of unstable equilibrium. It is therefore necessary to choose appropriate starting conditions. Thus suppose we wish to simulate the regrowth of zoned vegetation in an experimentally created bare area (for example, a cleared intertidal rock face, or an expanse of bare mud in a newly drained marsh). One must obviously put $m_{x0} = n_{x0} = 0$ for all $x > 0$ since colonization of a bare area must necessarily be by spores or seeds (that is, by individuals in the 0-year old age class). In the present example, simulations were begun with three different combinations of the two species. They were

$$V_0'(1) = (\ 150 \quad 0 \quad 50 \quad 0)\ ,$$

$$V_0'(2) = (\ 100 \quad 0 \quad 100 \quad 0)\ ,$$

and $$V_0'(3) = (\ 50 \quad 0 \quad 150 \quad 0)\ .$$

Intrinsically improbable founder populations were not used. Thus $V_0(1)$, in which M's were three times as numerous as N's, were not used at $G = 25$ (in the stable case) or at $G = 75$ (in the unstable case) for one would not expect M's to be more abundant than N's in close proximity to the zone where N is the invariable winner. Likewise $V_0(3)$ was not used at $G = 25$ (unstable) or $G = 75$ (stable).

What appeared to be the most useful results of the simulations are summarized in Table 2 (on the following page). The best diagnostic difference between the stable and unstable models was provided by the time taken for the percentage of 0-year old individuals in each species to fall from the initial 100% value to 70% (an arbitrarily chosen level) on its way to the ultimate stable value of 66.7%. These times are the values tabulated in Table 2. The unit of time

TABLE 2

TIMES TO "NEAR STABILITY" OF THE AGE DISTRIBUTIONS
IN POPULATIONS OF TWO COMPETING SPECIES

(The time unit is 2 years. The entries show the time at which the percentage of 0-year old individuals in each species first fell below 70% . For realizations of the unstable model the time is given, in parentheses, at which the total size of the losing population fell to one-tenth that of the winning population)

G	Initial vector	Unstable Model		Stable Model	
		M	N	M	N
	$V_0(1)$	2	>20 (3)	-	-
25	$V_0(2)$	3	>20 (7)	5	3
	$V_0(3)$	-	-	4	4
	$V_0(1)$	3	>20 (6)	4	3
50	$V_0(2)$	4	4 (21)	4	4
	$V_0(3)$	>20 (8)	3	3	4
	$V_0(1)$	-	-	4	4
75	$V_0(2)$	>20 (8)	3	3	5
	$V_0(3)$	>20 (3)	3	-	-

is 2 years. (The state of affairs after only one year was not considered; at the outset of a simulation slight "jogs" in age structure occur whose magnitude is very sensitive to changes in the composition of the founder populations.) In some cells of the table there is a number in parentheses after the main entry; this is the time (again in 2-year units) for the total abundance (both age classes) of the winning species to become ten times as great as that of the loser.

The most conspicuous qualitative result is that in a zone of unstable equilibrium the percentage of young individuals in the species destined to lose remained consistently high; whereas the age distribution of the winning species approached the stable age distribution fairly quickly.

DISCUSSION AND CONCLUSIONS

It would be easy to generate large numbers of realizations of the model assigning various combinations of numerical values to the parameters. But to do this in a vacuum, without some particular example in mind, would obviously be unrewarding since there are so many parameters for which arbitrary values have to be chosen in every case.

Ideally, the next step in this investigation would be to observe the reestablishment of zoned vegetation on experimentally denuded areas with strong environmental gradients, noticing especially how the age distributions of the colonizing species changed with time at different points on the gradient. Such experiments would have to be of several years' duration and the workers conducting them would undoubtedly modify the model in various ways while the experiment was in progress. Thus in the light of gradually accumulating evidence one might be able to make more realistic assumptions as to the dependence of the growth rates per individual (that is, $(1/M)dM/dt$ and $(1/N)dN/dt$) on M_t, N_t and G. The assumption so far made (in the continuous time form of the model), namely that these rates are a linear function of M_t , N_t and G is certainly reasonable in a small neighborhood of M_e and N_e and in a small neighborhood of the G-value at the zone boundary; (the word neighborhood is used here in its mathematical sense in regard to M and N and in both its mathematical and colloquial senses in regard to G). But it remains to discover how large, in any given case, a "small neighborhood" can be assumed to be.

Although experimental testing of the theory will take several years, observations on existing zoned plant communities should certainly provide some evidence as to the tenability of the model. In particular, it would be interesting to know whether, when colonizing species invade a bare area where there is an environmental gradient, the age distribution of each species varies along the gradient. The results given in Table 2, obtained from simulations of the model, suggest that at points on the gradient from which it will ultimately be excluded a species is likely to have a higher proportion of young individuals than occurs in a population of stable age distribution. Consequently, in a zone where it is fighting a losing battle (and is destined to be excluded) a species would be expected to have a disproportionately large number of young individuals compared with the proportion present in the zone where it is winning and to which it will ultimately be confined. It would clearly be worth while for ecologists interested in the zonation of vegetation to collect evidence on this point.

The model as it now stands gives a good qualitative description of seaweed zonation. Thus it explains, in a very satisfactory manner, why the zonation of intertidal Phaeophyta should be so clearcut on sheltered shores where algal growth presumably continues until a shortage of resources halts it; and so blurred on exposed shores where wave action prevents the algae ever becoming dense enough to be resource-limited (Lewis 1964). It also explains why hybrids (presumably competitively weaker than their parents) are normally found only at exposed sites where competition is comparatively mild (Burrows and Lodge 1951). By postulating the existence of unstable transition zones, the model disposes of the need to postulate mysterious "critical tide factors" (Doty 1946, Lewis 1964) to mark the various zone boundaries.

Finally it should be noticed that wherever an environmental gradient has caused plant communities to exhibit strong zonation with abrupt boundaries, nature has set up a field experiment for us. For, as a result of natural causes, the member species of a many-species community have become sequestered in such a way that each is fighting only two competitive battles, one on the "uphill" side and one on the "downhill" side of its zone; and at every zone boundary a simple two-species confrontation is going on.

ACKNOWLEDGMENTS

I have profited greatly from discussions with Dr. A.R.O. Chapman on zonation in seaweeds.

This work was funded by a grant from the National Research Council of Canada.

BIBLIOGRAPHY

Burrows, E.M. and Lodge,S. (1951). Autecology and the species problem in Fucus. *J. Marine Biol. Assoc. U.K.* 30: 161 - 175.

Doty, M. (1946). Critical tide factors that are correlated with the vertical distribution of marine algae and other organisms along the Pacific coast. *Ecology* 27: 315 - 327.

Keyfitz, N. (1968). Introduction to the mathematics of population. Addison-Wesley, Reading, Mass.

Leslie, P.H. (1945). The use of matrices in certain population mathematics. *Biometrika* 33: 183 - 212.

Levin, S.A. (1970). Community equilibria and stability, and an extension of the competitive exclusion principle. *Amer. Natur.*,104:413-423.

Levins, R. (1968). Evolution in changing environments. Princeton University Press, Princeton, N.J.

Lewis, J.R. (1964). The ecology of rocky shores. English Universities Press, London.

MacArthur , R.H. (1970). Species packing and competitive equilibrium for many species. *Theoret. Pop. Biol.* 1: 1 - 11.

May, R.M. (1973). On relationships among various types of population models. *Amer. Natur.* 107: 46 - 57.

Strobeck, C. (1973). *N* species competition. *Ecology* 54: 650 - 654.

Vandermeer, J.H. (1970). The community matrix and the number of species in a community. *Amer. Natur.* 104: 73 - 83.

MORPHOGENESIS IN SAND DOLLAR EMBRYOS

John W. Prothero
School of Medicine, Department of Biological Structure
University of Washington, Seattle.

ABSRACT

The early development (blastulation) of sand dollar embryos will be illustrated with a color movie film. Methods used to obtain a quantitative three-dimensional description of the process in single living animals will be reviewed. These include: (1) photomicrography of serial optical sections, (2) digitization of photographic contours and (3) computation of spatial parameters. The results of a preliminary study will be reported and some problems for the future outlined. A paper describing the working has been submitted to the Journal of Microscopy.

ECOLOGICAL AND ECONOMIC MARKETS

David J. Rapport
Department of Biological Sciences
Simon Fraser University
Burnaby, British Columbia

James E. Turner
Department of Mathematics
McGill University
Montréal, Québec

1. Introduction

The literature of economics contains many examples of analogies and transfers of concepts between biological and economic systems. Boulding (1972) outlines three major areas where these parallels are found: the theory of the organism and its behavior, the theory of the ecosystem (the interaction of populations of different organisms), and the theory of evolution (the succession of ecosystems through time). Models of the economics of species behavior have been developed recently by a number of authors (Rapport, 1971; Schoener, 1971; MacArthur, 1972; Turner and Rapport, this volume). Generally, these models attempt to explain the foraging behavior of predators in terms of some measure of welfare such as net assimilated energy or reproductive success. In this paper, a model of population interactions is developed using concepts similar to those used in the analysis of economic markets. Dynamical models of population interactions such as the Lotka-Volterra system will be seen to be similar to dynamical models of commodity prices and quantities in economic markets.

2. Harvest and Yield Functions

Consider an ecosystem of m interacting populations with population vector $\underline{n}(t) = (n_1(t), n_2(t), \ldots, n_m(t))$. The dynamics of each population is determined by a knowledge of two functions. The yield function $f_i(\underline{n})$ of the $_i$th species is defined to be the difference between the fecundity rate and the non-predator mortality rate. The harvest function $h_i(\underline{n})$ of the $_i$th species is the predator mortality rate; i.e. the rate at which the $_i$th species is consumed by all predators. In general, the harvest and yield functions depend on the densities of all populations in the ecosystem.

The dynamics of the ecological community is given by the system of first order equations

$$\frac{dn_i}{dt} = f_i(\underline{n}) - h_i(\underline{n}) \qquad i = 1,2,\ldots,m \, . \tag{1}$$

In principle, knowledge of the harvest and yield functions determines the time development of the ecosystem. The equilibria of the community are determined by the m simultaneous equations $f_i(\underline{n}) = h_i(\underline{n})$. Sufficient conditions for the stability of an equilibrium are, at the equilibrium,

$$\frac{\partial f_i}{\partial n_i} < \frac{\partial h_i}{\partial n_i} \text{ and } \frac{\partial f_i}{\partial n_j} > \frac{\partial h_i}{\partial n_j} \qquad i \neq j \, . \tag{2}$$

These sufficient conditions for stability imply that an increase in the $_j$th population from its equilibrium value increases the yield of other populations more than the harvest. (In economic terms, these conditions indicate that all populations are gross substitutes as prey.) These conditions can be derived by linearizing system *(1)* near the equilibrium.

The Lotka-Volterra model of community interactions

$$\frac{dn_i}{dt} = n_i(r_i - \sum_{j=i}^{m} a_{ij}n_j) \tag{3}$$

can be interpreted as a special case of system *(1)* in a number of ways. The simplest way is to partition the components of the community matrix $A = (a_{ij})$ to separate predator and non-predator mortality factors. If such a partition is possible, the Lotka-Volterra system can be written

$$\frac{dn_i}{dt} = n_i(r_i - \sum_{j=i}^{m} c_{ij}n_j) - n_i \sum_{j=i}^{m} h_{ij}n_j \,. \tag{4}$$

The components of $C = (c_{ij})$ measure non-predator mortality factors such as crowding and the components of $H = (h_{ij})$ measure mortality due to predation.

Complex food webs can be simulated by realistic assumptions of the possible forms of the harvest and yield functions in system *(1)*. One purpose of constructing such simulation models would be to understand the relative importance of predator and non-predator mortality in ecological communities.

3. The Ecological Market

The ecological market may be defined as the balancing process of all harvest and yield functions in a natural community. There is a close resemblance between this process and the balancing of supply and demand rates in economic markets. In an economic market, producers and consumers interact to determine equilibrium prices and quantities. Typical supply and demand rates for an economic good as functions of price are shown in Fig. 1a. In an ecological market, prey and predators interact to determine equilibrium prey densities and harvest rates. In Fig. 1b, yield and harvest functions are drawn for a particular prey as functions of prey density.

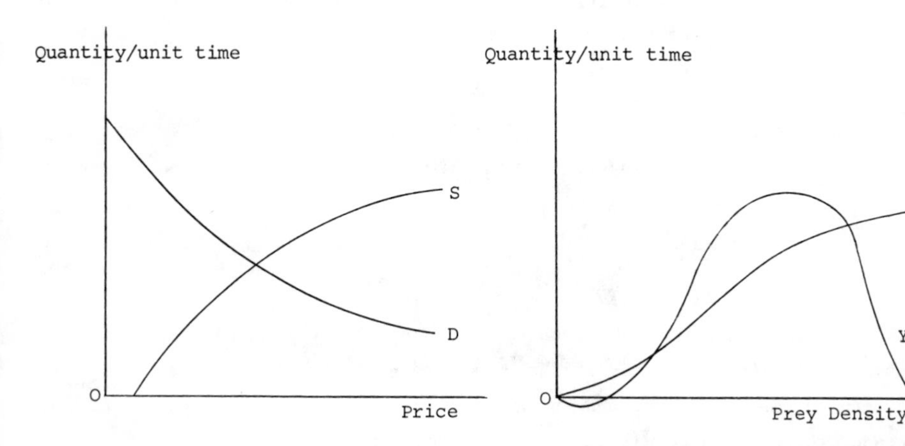

Figure 1a: Supply and Demand

Figure 1b: Yield and Harvest

As in Fig. 1b, consider the harvest and yield functions for a given prey species and assume, for the moment, that all other populations are constant. Then the (partial) equilibrium densities of this prey are determined by the intersections of the harvest and yield functions. As shown in the diagram, there may be more than one equilibrium prey density. A corresponding phenomenon has been noted in some economic markets, particularly in labor and foreign exchange markets.

The positions of these equilibria will change in response to changes in predator densities and in the densities of alternative prey. The equilibria of economic markets are also partial equilibria since, in Fig. 1a, it is assumed that the prices of all other commodities are fixed. The prices of alternative commodities change in response to a change in price of the given commodity. This in turn changes the supply and demand rates for this good, and therefore the equilibrium price may be altered (Arrow and Hahn, 1971). The graphical analysis of the complex adjustment process has proved to be very useful in gaining a qualitative understanding of economic markets. In a similar way, qualitative information about natural communities can be obtained by considering the graphical model of harvest and yield functions. For example, an increase in the predator populations will increase the harvest function of the given prey and reduce the equilibrium prey densities. Increases in the food resources available to the prey population will increase the yield function of this prey and, in turn, increase the equilibrium prey densities. Many other special cases can be illustrated very simply by the graphical model.

4. Market Mechanisms in Economics and Ecology

Both economic and ecological markets can be viewed as mechanisms for allocating limited resources to alternative activities. The law of supply and demand in economic markets is the rule: raise price if total demand exceeds total supply and lower price if total supply exceeds total demand. Walras (1874) developed the first dynamical model of market equilibrium. If there are m commodities in an economic market with prices $\underline{p}(t) = (p_1(t), p_2(t), \ldots, p_m(t))$, then the approach to market equilibrium can be described by the system of first order equations

$$\frac{dp_i}{dt} = g_i(E_i(\underline{p}(t))) \tag{5}$$

where $E_i(p(t))$ is the excess demand for the $_i$th good and g_i is a monotone increasing function of its argument satisfying $g_i(0) = 0$. The stability of the linear "tatonnement" system

$$\frac{dp_i}{dt} = E_i(\underline{p}(t)) = D_i(\underline{p}(t)) - S_i(\underline{p}(t)) \tag{6}$$

is discussed by Intriligator (1971).

The dynamics of an ecological community of system *(1)* corresponds very closely to the linear tatonnement system *(6)*. The growth rate of the $_i$th population is the difference between the yield rate and the harvest rate. This is the negative of the "excess demand" for the $_i$th population.

The Lotka-Volterra equations *(3)* have been interpreted as a special case of the general harvest and yield model *(1)*. This gives some insight into the basic similarity of the dynamics of price in economic markets and the dynamics of prey density in natural communities.

5. Discussion

Boulding (1972) has made the suggestion to compare the work of Walras on the general equilibrium of the price system with the work of Volterra on the dynamic interactions of populations. In this paper, we have developed a model of the dynamics of natural communities based on an analogy between economic and ecological markets. Resource supplies are brought into balance with resource demands by a complex tatonnement or groping process. The essential requirement for achieving a balance is a negative feedback response when supplies and demands are not equal. In economic systems, relative prices are responsive to excess demand while, in ecological systems, relative densities of prey are responsive to disequilibrium.

We have indicated, in very general terms, how the resource allocation process in economic markets is analogous to resource allocation in natural communities. In a second paper in this volume, we outline a model of ecological population growth and competition in economic terms. Experimental and theoretical work is in progress to investigate the similarities and differences of allocation mechanisms in economics and ecology.

ACKNOWLEDGEMENTS

The suggestions and encouragement of Dr. A. L. Turnbull were very helpful during the development of these ideas. This work was supported by a Canada Council I. W. Killam Senior Research Scholarship.

REFERENCES

Arrow, K. J. and Hahn, F. H. General Competitive Analysis. Holden-Day, San Francisco (1971).

Boulding, K. E. Economics as a Not Very Biological Science. IN J. A. Behnke (ed.), Challenging Biological Problems. Oxford University Press, New York (1972).

Intriligator, M. D. Mathematical Optimization and Economic Theory. Prentice-Hall, Englewood Cliffs (1971).

MacArthur, R. H. Geographical Ecology. Harper and Row, New York (1972).

Rapport, D. J. An Optimization Model of Food Selection. Amer. Natur. 105, 575-587 (1971).

Schoener, T. W. Theory of Feeding Strategies. Ann. Rev. Ecol. Systematics 2, 369-404 (1971).

Turner, J. E. and Rapport, D. J. An Economic Model of Population Growth and Competition in Natural Communities, this volume.

Walras, L. Elements of Pure Economics. English translation of the definitive edition by Wm. Jaffé, Allen and Unwin, London (1954).

SOME STOCHASTIC GROWTH PROCESSES

Charles E. Smith and H.C.Tuckwell

Department of Biophysics and Theoretical Biology,

University of Chicago

Abstract. The growth of non-saturating and saturating populations is modeled by a general kind of stochastic differential equation. The transition density functions of the solutions of these equations, obtained using the Stratonovich stochastic integral, are obtained in closed form. Moments, first passage time probability densities and probabilities of extinction can be found explicitly in a number of cases. Specifically considered are Malthusian growth, a general non-saturating process, a general saturating process which contains the Pearl-Verhulst model as a special case, and Gompertzian growth. This last-named process is examined with a view to the stochastic modeling of large populations of tumor cells.

1. INTRODUCTION

Recently there has been much interest in problems connected with the effects of random influences on the growth of populations. Goel et al. (1971) and Montroll (1972) have studied populations whose size $N(t)$ at time t satisfies a stochastic differential equation of the form

$$dN(t) = kN(t)G(N(t)/K)dt + N(t)dw(t), \tag{1}$$

where $G(\cdot)$ is a saturation function, K is the (constant) saturation level, k is a growth rate parameter and $w(t)$ is a Wiener process with zero mean. The random term $N(t)dw(t)$ might reflect fluctuations in the physical features of the environment or the effects of other species. In the latter case, n equations of type (1) can be used as a (degenerate) model of the Lotka-Volterra type where the number of species is n.

There seems, however, to be a basic problem in the approach of the above-mentioned references in that when the random term is added to the deterministic equation, which represents a saturating (bounded) process, the resulting random process $N(t)$ takes values in $[0,\infty)$. Hence the "saturating" feature is, in the strict sense, lost.

In this paper we will consider populations whose sizes satisfy stochastic equations of the type

$$dN(t) = h(N(t))dw(t), \tag{2}$$

where the function h(N) may or may not be of the saturating form. Such equations imply that a growth rate parameter will be taken to be a Gaussian white noise, which may be looked upon as the formal derivative, dw(t)/dt, of a Wiener process. Furthermore, we will not necessarily assume that w(t) has a zero mean.

It is known that, provided h(N) satisfies certain regularity conditions (Doob,1953), equation (2) has the solution

$$N(t) = N(0) + \int_0^t h(N(t'))dw(t'), \tag{3}$$

which is a continuous Markov process. That is, N(t) is a diffusion process whose transition probability density function(p.d.f.), satisfies the Fokker-Planck equation

$$\frac{\partial p}{\partial t} = -\frac{\partial}{\partial N}[M(N)p] + \frac{V}{2}\frac{\partial^2}{\partial N^2}[h(N)^2 p], \tag{4}$$

where

$$p(N,t|N_o)dN = \Pr[\, N < N(t) \leq N+dN |\, N(0) = N_o], \tag{5}$$

V being the variance parameter of w(t).

Ito and Stratonovich Calculi.

The first infinitesimal moment, M(N), depends upon the way in which the stochastic integral in (3) is defined. Using obvious subscripts to denote whether the Ito or Stratonovich definition is being employed (Jaswinski,1970), we have

$$M_I(N) = mh(N), \tag{6}$$

$$M_S(N) = mh(N) + \frac{V}{2}h(N)h'(N), \tag{7}$$

where m is the mean value of dw(t)/dt and the prime denotes differentiation. The additional term in the Stratonovich first moment indicates that h(N(t)) and dw(t)/dt have the covariance,

$$\mathrm{COV}_S[h(N(t)),dw(t)/dt] = (V/2)E[h(N(t))h'(N(t))], \qquad (8)$$

where E denotes mathematical expectation.

It has been pointed out that the Stratonovich calculus is the better one to adopt if the solution of a stochastic equation represents a "physical" random process (Gray & Caughey,1965), and this argument has found mathematical support in a theorem of Wong & Zakai (1965). This would imply that when modeling the stochastic behaviour of populations with equations such as (1) or (2), one should employ the Stratonovich definition of the stochastic integral, though some authors have expressed some skepticism concerning the fact that there is a "choice" of calculi (see,e.g.,Mortensen,1969).

A distinct advantage in using the Stratonovich calculus is that regular calculus rules are preserved. This means that one can often find explicit (closed-form) solutions of the Fokker-Planck equation by simply transforming the original stochastic differential equation to that of a Wiener process (Lax,1966; Tuckwell,1974). Similarly, in some cases, the first passage time p.d.f. may also be written down in closed form (see section 3).

2. NON-SATURATING GROWTH PROCESSES

During certain stages of growth, many populations of cells or organisms evolve according to deterministic equations of the general form

$$\frac{dN}{dt} = rN^b, \quad 0 < b \leq 1, \qquad (9)$$

where r is the growth rate parameter which contains the net effect of the processes of birth, death, immigration and emigration. When r is regarded as a random process, r(t), its fluctuations represent the influences of the stochastic behaviour of environmental variables such as nutrient or food supply, climatic conditions, etc. To obtain some insight into how these fluctuations affect the growth of a population we may regard r(t) as a Gaussian white noise, with the properties

$$E[r(t)] = m, \tag{10}$$

$$COV[r(t_1), r(t_2)] = V\,\delta(t_1 - t_2). \tag{11}$$

Under these assumptions equation (9) is of the form of (2) and the transition p.d.f. of N(t) thus provides a complete description of the process.

If we put b = 1 in (9) we obtain the usual Malthusian model which has been treated from a mathematical viewpoint by Gray & Caughey (1965) and in studies of population growth by Goel et al. (1971) and by Capocelli & Ricciardi (1973). We will emphasize and discuss the difference in the results obtained when the Ito and Stratonovich versions of the first infinitesimal moment are employed in equation (4).

The case $0 < b < 1$, which represents a general non-saturating process, has not been treated before. Deterministic solutions take the form

$$N(t) = [\,(1 - b)rt + N_o^{1-b}]^{(1/(1-b))} \tag{12}$$

and biological motivation is obtained from experimental results on certain tumor systems (Mendelsohn & Dethlefsen,1968; Lala,1971). It turns out that the random process N(t) for this range of values of b has features which clearly distinguish it from the Malthusian case.

2.1 Malthusian Growth (b = 1).

The transformed process,

$$Y(N) = \int^{N} [h(N')]^{-1} dN' , \tag{13}$$

which satisfies the equation of the Wiener process, in this case takes values in $(-\infty,\infty)$. The transition p.d.f. can be found immediately in closed form for both Ito and Stratonovich approaches, which means that a direct comparison of the results is possible. The following expressions are readily obtained.

Ito Calculus.

$$p_I(N,t|N_o) = \frac{N^{-1}}{\sqrt{2\pi Vt}} \exp\left[\frac{-[\log(N/N_o) - (m - V/2)t]^2}{2Vt}\right], \quad (14)$$

$$E_I(N(t)|N_o) = N_o \exp(mt), \quad (15)$$

$$VAR_I(N(t)|N_o) = N_o^2 \exp(2mt)[\exp(Vt) - 1], \quad (16)$$

$$P_I(m,V) = \begin{cases} 0 \ , & \text{if } m > V/2, \\ 1/2, & \text{if } m = V/2, \\ 1 \ , & \text{if } m < V/2. \end{cases} \quad (17)$$

Here VAR denotes variance and P is the probability of ultimate extinction, which is in this case defined to be

$$\lim_{t \to \infty} \Pr[\, N(t) \leq \Delta \mid N(0) = N_o], \ 0 < \Delta < \infty. \quad (18)$$

This definition must be employed here because the boundaries of the diffusion process ($N = 0$ and $N = \infty$) are natural (see Feller,1952, for method of classification) which means that they can never be reached in a finite time. With $N = 0$ an accessible boundary, a different definition is employed (see section 2.2).

Stratonovich Calculus.

The quantities p_S, E_S, VAR_S, and P_S are obtained from the corresponding expressions (14) - (17) by simply replacing m by $m + V/2$. Furthermore, whereas

$$COV_I[N(t),r(t)] = 0, \quad (19)$$

we have, from (8),

$$COV_S[N(t),r(t)] = (N_o V/2)\exp((m + V/2)t). \quad (20)$$

In considering the asymptotic behaviour of the above quantities as $t \to \infty$, we must distinguish seven cases. For the Ito approach:

(i) $-\infty < m < -V/2$: $E(N) \to 0$; $VAR(N) \to 0$; $P = 1$.

(ii) $m = -V/2$: $E(N) \to 0$; $VAR(N) \to N_o^2$; $P = 1$.

(iii) $-V/2 < m < 0$: $E(N) \to 0$; $VAR(N) \to \infty$; $P = 1$.

(iv) $m = 0$: $E(N) = N_o$; $VAR(N) \to \infty$; $P = 1$.

(v) $0 < m < V/2$: $E(N) \to \infty$; $VAR(N) \to \infty$; $P = 1$.

(vi) $m = V/2$: $E(N) \to \infty$; $VAR(N) \to \infty$; $P = 1/2$.

(vii) $V/2 < m < \infty$: $E(N) \to \infty$; $VAR(N) \to \infty$; $P = 0$.

The results for the Stratonovich approach are again obtained by replacing m by m + V/2 in the inequalities (i) - (vii).

Lewontin & Cohen (1969) have already pointed out that it seems "paradoxical" for the expected population size to become infinite as $t \to \infty$, while the extinction probability is unity. This, and other "paradoxes", occur in both calculi. For example, when we examine the expected overall growth rate, we find

$$E_I[dN(t)/dt] = E_I[r(t)]\,E_I[N(t)] , \tag{21}$$

and $E_I[N(t)] > 0$ for all t. Thus the expected rate of change of the population size has the sign of m. Hence it is possible, if $0 < m < V/2$, for the expected value of dN/dt to be always positive even though the probability of extinction is one. Paradoxes such as these do not lead to any intuitive support for a choice of calculi,but it is pointed out that the Stratonovich calculus does yield the somewhat appealing result that extinction is certain only if the average value of the growth rate parameter,r(t),is negative.

2.2 General Non-Saturating Process ($0 < b < 1$).

In this case the transformation

$$Y(N) = \int^{N} N'^{-b}dN' = N^{1-b}/(1-b), \tag{22}$$

leads to a process Y(t) which is a Wiener process in $[0,\infty)$. The boundary at the origin is regular (Feller,1952) so that boundary conditions may be imposed upon the solutions of (4). Clearly N = 0 is absorbing, so that

$$p(0,t|N_o) = 0 \tag{23}$$

is the appropriate constraint. Employing the Stratonovich calculus (as we shall throughout the rest of this paper), the Fokker-Planck equation becomes

$$\frac{\partial p}{\partial t} = -\frac{\partial}{\partial N}[N^{b}(m + \frac{bVN^{b-1}}{2})p] + \frac{V}{2}\frac{\partial^2}{\partial N^2}[N^{2b}p]. \tag{24}$$

The solution of this equation (Green's function) satisfying (23) is

$$p(N,t|N_o) = \frac{N^{-b}}{\sqrt{2\pi Vt}}\left(\exp\left[\frac{-[Y(N) - Y(N_o) - mt]^2}{2Vt}\right] - \exp\left[\frac{-2mN_o^{1-b}}{V(1-b)}\right]\exp\left[\frac{-[Y(N) + Y(N_o) - mt]^2}{2Vt}\right]\right), \tag{25}$$

where $Y(N)$ is defined in equation (22).

Since zero population level is represented by an accessible barrier, in this case we can define the probability of extinction at t as

$$P(t|N_o) = 1 - \int_0^\infty p(N,t|N_o)dN$$

$$= \Phi\left(\frac{(b-1)^{-1}N_o^{1-b} - mt}{\sqrt{Vt}}\right) + \exp\left[\frac{-2mN_o^{1-b}}{V(1-b)}\right] \times \left[1 - \Phi\left(\frac{-(b-1)^{-1}N_o^{1-b} - mt}{\sqrt{Vt}}\right)\right], \tag{26}$$

where $\Phi(\cdot)$ is the normal distribution function. This definition is the same as that employed by Feller (1951). Taking the limit as $t \to \infty$, we find that the probabilities of ultimate extinction are

$$P(\infty|N_o) = \begin{cases} 1, \text{ if } m \le 0, \\ (1/2)\exp[-2mN_o^{1-b}/((1-b)V)], \text{ if } m > 0. \end{cases} \tag{27}$$

From this result we see again that the use of the Stratonovich calculus leads to the prediction that $m = 0$ is the critical value associated with certain extinction. Furthermore, if survival is possible ($m>0$), then the probability of ultimate extinction is enhanced by increasing the variance parameter of the "intrinsic" growth rate. It might therefore be concluded that, the more uncertain the environment, the less is the chance of survival for given values of N_o , b and m .

3. SATURATING GROWTH PROCESSES

While some stages of growth may be governed by equations such as (9), growth in the latter stages is usually restricted by such factors as physical boundaries, competition, predation, limitations in food or nutrient supply etc. The growth rate then becomes "density-dependent" and diminishes as the carrying capacity (saturation level) is approached.

3.1 General Saturating Processes.

Montroll (1972) has introduced (c.f. equation (1)) the family of saturation functions

$$G_\alpha(x) = \frac{1 - x^\alpha}{\alpha} . \tag{28}$$

The case $\alpha = 1$ yields the well known Pearl-Verhulst logistic process, whereas $\alpha = 0$ corresponds to the Gompertzian growth process (see section 3.2).

Several stochastic treatments of the logistic process have appeared, some treating the intrinsic growth rate as a random process (Levins,1969; Tuckwell,1974), others regarding the carrying capacity as a fluctuating parameter (Levins,1969; May,1973) or adding a noise term as in equation (1) (Goel *et al.*,1971).

An alternative form of saturation function is defined if we write

$$\frac{dN}{dt} = rN[1 - (N/K)]^\nu , \quad \nu > 0. \tag{29}$$

If $\nu = 1$ we recover the standard logistic process, whereas $\nu = 0$ gives rise to the non-saturating Malthusian law. Suppose we let r in (29) be a Gaussian white noise, as in section 2; then the resulting stochastic equation is of the same form as (2). For positive integer values of ν the transformed process

$$Y(N) = \log[N/(K - N)] + \sum_{j=1}^{n-1} \binom{n-1}{j} j^{-1} \left[\frac{N}{K-N}\right]^j , n=1,2,.. \tag{30}$$

is a Wiener process in $(-\infty,\infty)$, and since (30) is a strictly monotonic function the transition p.d.f. of $N(t)$ can be obtained in closed form. Hence the conditional moments of $N(t)$ can be found by numerical integration. It can be seen that the probability of ultimate extinction, defined as in (18) because the boundaries $N=0$ and $N=K$ are natural, is unity whenever $m < 0$. As with all the processes so far considered, $N(t)$ does not have a stationary distribution as $t \to \infty$.

3.2 Gompertzian Growth and Tumor Biology

The Pearl-Verhulst logistic process characterizes, to a close approximation, the growth of many populations of organisms (bacterial, animal, human etc.) but many systems of cells in complex organisms are apparently much more accurately described by the Gompertz equation,

$$\frac{dN}{dt} = rN\log(K/N), \quad 0 < N(0) < K, \tag{31}$$

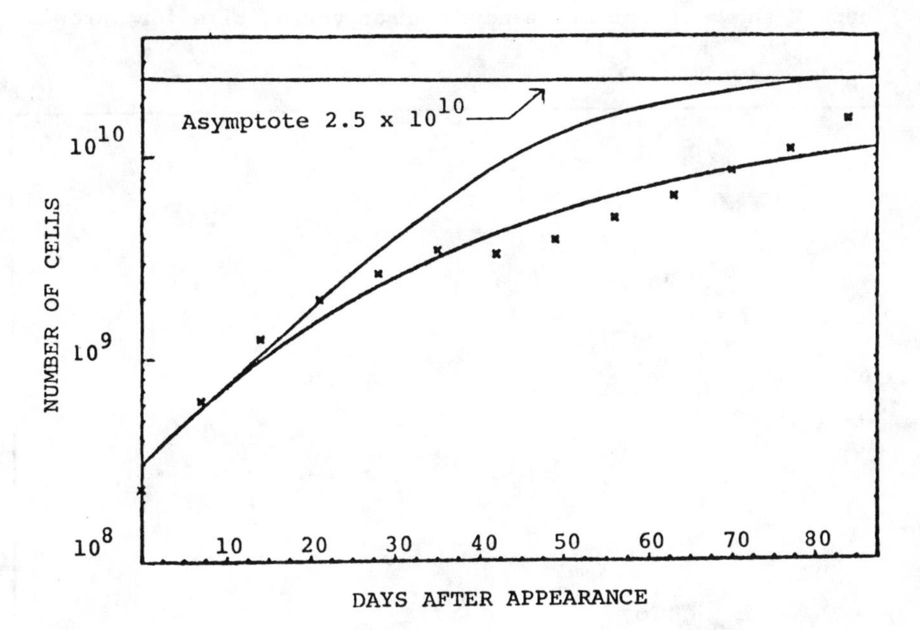

Figure 1. Mean size data (x) and best fit Gompertz curve for DMBA tumors (after Simpson-Herren & Lloyd,1970). The upper line is a logistic curve (see text).

where K is the saturation level. Solutions of this equation,

$$N(t) = K\exp[\log(N_o/K)e^{-rt}], \tag{32}$$

(or equivalent forms) were found to fit well the experimental growth data for many embryos, organs and tumors (Laird, 1964,1965,1969). Subsequently, in comprehensive studies, the Gompertz curve has been used to fit the mean size data of a variety of solid and ascites tumors (Simpson-Herren & Lloyd, 1970 and references therein). Figure 1 shows a typical Gompertzian growth curve. The data is for mammary adenocarcinomas induced by intra-gastric doses of DMBA (7,12-dimethylbenz(a) anthracene) in female Sprague-Dawley rats. Measurements were made (on growing tumors only) in two directions with calipers and the volumes calculated from an approximate formula for a prolate spheroid. For comparison, in the same figure, the (standard) logistic curve has been drawn with the same values of N_o, K, and r.

Measurements of the sizes of large numbers of tumors of the same age and type,in animals of the same species, show considerable variability. Figure 2 shows a plot of standard error versus time for (growing) DMBA tumors.

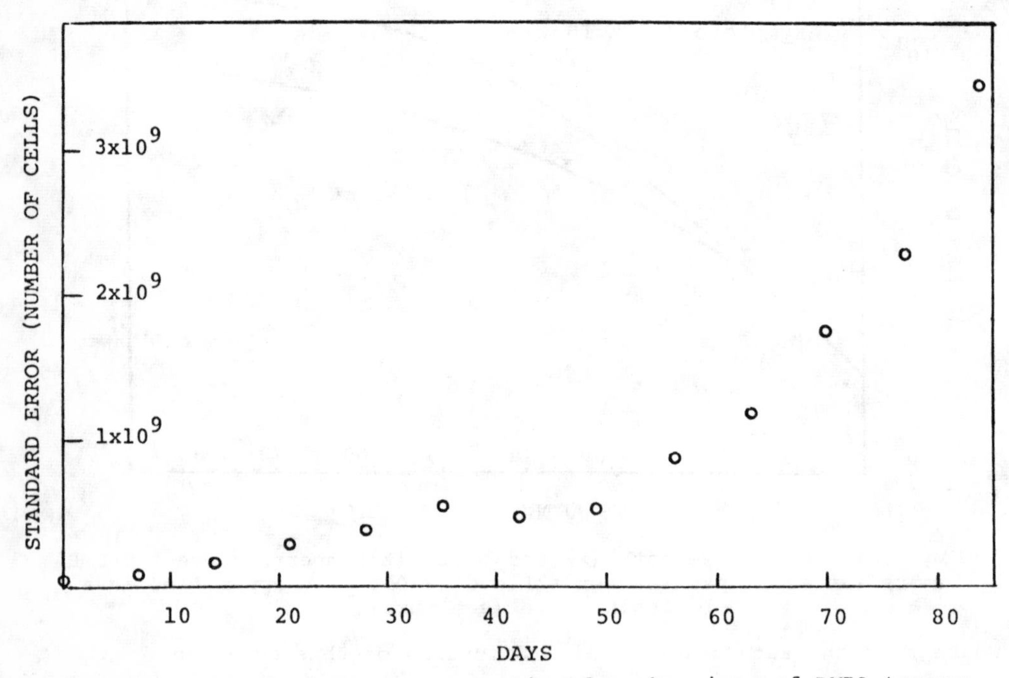

Figure 2. Standard error versus time for the sizes of DMBA tumors (data courtesy of Simpson-Herren, 1973).

Contributions to the variability are expected to come from differences in physiology, anatomy and host defense mechanisms from animal to animal. Furthermore, for cells within a given tumor, labelled mitosis techniques have shown that the cell cycle time has a distribution of values, and this, together with the growth fraction and rate of cell loss (see Lala, 1971, for a review of tumor cell population kinetics) affects the overall growth rate. These quantities will vary from animal to animal and depend upon such things as degree of vascularization and the spatial distribution of proliferating and non-proliferating cells (Burton,1966; Greenspan,1972) which in turn affect the oxygen and nutrient supply.

Further evidence of a stochastic element in tumor growth is manifested in the "surviving fraction" data, which represents the fraction, F(t), of host animals still alive at time t. An example of the variation in time of this quantity is shown in Figure 3. The nature of the data indicates that there is an approximate critical lethal size, reached at various times by tumors in different hosts, or that there are

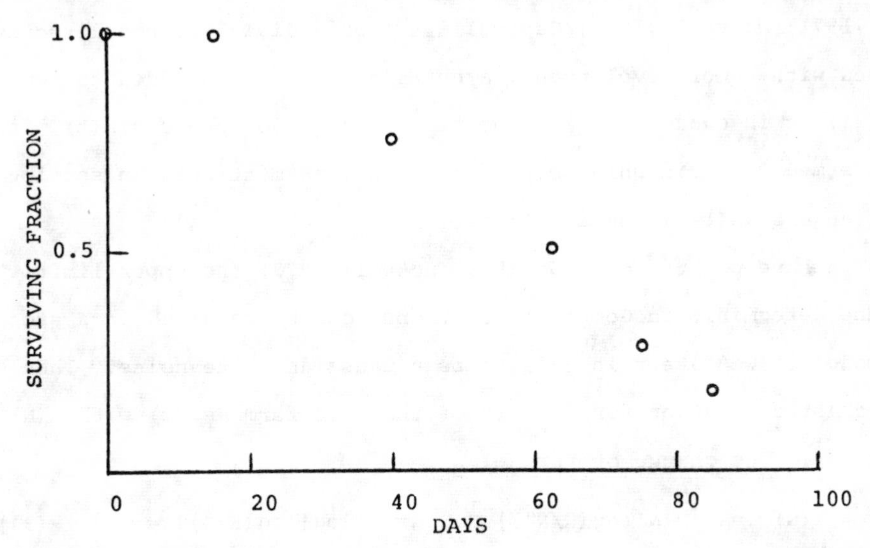

Figure 3. Surviving fraction data for DMBA tumors (from Simpson-Herren & Lloyd, 1970)

variations in the lethal size from animal to animal, or both of these factors may contribute to the variations in the time of death. If the first factor is predominant, then the surviving fraction at time t can be estimated from the probability density, $f(N^*,t|N_o)$, of the time of first passage of $N(t)$, to the critical size N^*, given an initial size N_o, through the relation

$$F(t) = 1 - \int_0^t f(N^*,t'|N_o)dt', \quad 0 < N_o < N^* < K, \tag{33}$$

where the random process $N(t)$ represents the number of tumor cells.

Of those studies which incorporate a stochastic approach to tumor growth, some have focused upon carcinogenesis, while many others have considered the relationship between cellular kinetic parameters and growth rate and the response of these quantities to chemotherapy. Recent reviews of such studies may be found in Aroesty et al. (1973), Iosifescu & Tautu (1973) and Smith (1972). Little attention, however, has been given to probabilistic descriptions of the overall growth of large populations of tumor cells.

3.3 Randomized Gompertzian Growth.

In previous stochastic models related to Gompertzian growth (Goel et al. ,1971; Montroll,1972; Capocelli & Ricciardi,1974), not directly concerned with tumor development, a noise term has been added to equation (31) as in equation (1). The resulting random process, $N(t)$, is a transformed Ornstein-Uhlenbeck process and, as mentioned in section 1, the range of $N(t)$ is semi-infinite.

As we have pointed out (Smith & Tuckwell,1973) the upper limit, K, of the deterministic Gompertz solutions, can be retained in a stochastic model if we take r in (31) to be a Gaussian white noise. Then the stochastic equation for $N(t)$ is of the same form as (2) with $h(N) = N\log(K/N)$. The change of variables

$$Y(N) = \int^N [N'\log(K/N')]^{-1}dN' = -\log[\log(K/N)], \tag{34}$$

results in a process $Y(t)$ which is a Wiener process in $(-\infty,\infty)$. Thus the transition p.d.f. of $N(t)$ is given by

$$p(N,t|N_o) = \frac{[N\log(K/N)]^{-1}}{\sqrt{2\pi Vt}} \exp\left[\frac{-[Y(N) - Y(N_o) - mt]^2}{2Vt}\right], \quad (35)$$

where Y(N) is now defined as in (34). The boundaries (N = 0 and N = K) are again natural and no (proper) stationary distribution exists as $t \to \infty$.

Analytic expressions for the mean and variance of N(t) appear difficult to obtain, but these quantities may be found numerically using (35). Nevertheless, it is possible to find the mean and variance of logN(t) analytically:

$$E[\log N(t)|N(0) = N_o] = \log K - \log(K/N_o)\exp[(\frac{V}{2} - m)t], \quad (36)$$

$$VAR[\log N(t)|N(0) = N_o] = [\log(K/N_o)]^2\exp[(\frac{V}{2} - m)t].$$
$$.[\exp(Vt) - 1]. \quad (37)$$

The similarity between these expressions and those for the corresponding quantities in randomized Malthusian growth, is explained by noting that Z(t) = logN(t) satisfies the simple equation

$$\frac{dZ}{dt} = r(t)[\log K - Z]. \quad (38)$$

The transition p.d.f. of Z(t) can thus be obtained when either the Stratonovich or Ito calculus is used. However, the Fokker-Planck equation for the transition p.d.f. of N(t) (obtained above for the Stratonovich approach) with the Ito version of the first infinitesimal moment, does not appear to admit a ready analytic solution.

The first passage time p.d.f., for a final value N* and initial value N_o, of the restricted random Gompertzian growth process we have considered, can be obtained in closed form from that of the Wiener process (Cox & Miller,1965) because the function defined in (34) is strictly monotonic. Thus we have

$$f(N^*,t|N_o) = \frac{[Y(N^*) - Y(N_o)]}{\sqrt{2\pi Vt^3}} \exp\left[\frac{-[Y(N^*)-Y(N_o)-mt]^2}{2Vt}\right], \quad (39)$$

where it has been assumed that m is positive and Y(N) is again as in equation (34).

It is considered that treating equation (31) as a stochastic differential equation,with the growth rate parameter r a random process, provides a useful method of representing the effects of random influences on the growth of large populations of tumor cells. As a first approximation, one could take r to be a white noise as in the above treatment, and this approach is being investigated. Numerical calculations of the moments of N(t) and a comparison with experimental data will be published in the near future, together with a discussion of the problems of parameter estimation and the incorporation of the "microscopic" transition probabilities into a single randomized growth rate.

ACKNOWLEDGEMENTS

The authors wish to thank Dr. L. Simpson-Herren for providing statistical data on various tumors, and Dr. S .Coyne for useful comments on the manuscript. The financial support of an N.I.H. training grant and an Alfred P. Sloan Foundation Fellowship is acknowledged.

REFERENCES

Aroesty, J., Lincoln, T., Shapiro, N. & Boccia, G. Math. Biosci. 17, 243-300, (1973).

Burton, A.C. Growth 30, 157-176, (1966).

Capocelli, R.M. & Ricciardi, L.M. Theor. Pop. Biol. 4, (in press).

Capocelli, R.M. & Ricciardi, L.M. Theor. Pop. Biol. 5, (to appear).

Cox, D.R. & Miller, H.D. The Theory of Stochastic Processes. New York: Wiley, (1965).

Doob, J.L. Stochastic Processes. New York: Wiley,(1953).

Feller, W. Proc. 2nd Berkeley Symp. Math. Stat. Prob., 227-246, Berkeley: Univ. California Press, (1951).

Feller, W. Ann. Math. 55, 468-519, (1952).

Goel, N.S., Maitra, S.C. & Montroll, E.W. Rev. Mod. Phys. 43, 231-276, (1971).

Gray, A.H. & Caughey, T.K. J. Math. & Phys. 44, 288-296, (1965).

Greenspan, H.P. Studies in Applied Math. 51, 4, 317-340, M.I.T., (1972).

Iosifescu, M. & Tautu, P. Stochastic Processes and Applications in Biology & Medicine. Berlin,Heidelberg,New York: Springer-Verlag, (1973).

Jaswinski, A.H. Stochastic Processes & Filtering Theory. New York, London: Academic Press, (1970).

Laird, A.K. Brit. J. Cancer 18, 490-502, (1964)

Laird, A.K. Brit. J. Cancer 19, 278-291, (1965).

Laird, A.K. In Nat. Cancer Inst. Monograph 30, "Human Tumor Cell Kinetics", 15-28, (1969).

Lala, P.K. In Methods in Cancer Research 6, (H. Busch, Ed.), 3-95. New York, London: Academic Press, (1971).

Lax, M. Rev. Mod. Phys. 38, 541-566, (1966).

Levins, R. Proc. Nat. Acad. Sci. 62, 1061-1065, (1969).

Lewontin, R.C. & Cohen, D. Proc. Nat. Acad. Sci. 62, 1056-1060, (1969).

May, R.M. Stability & Complexity in Model Ecosystems. Princeton: Princeton Univ. Press, (1973).

Mendelsohn, M.L. & Dethlefsen, L.A. In The Proliferation & Spread of Neoplastic Cells. (Univ. Texas M.D. Anderson Hosp. & Tumor Inst., Ed.),197-212, Baltimore: The Williams & Wilkins Co., (1968).

Montroll, E.W. In Some Mathematical Questions in Biology.III. (J.D. Cowan, Ed.), 99-143, Providence,Rhode Island: The Amer. Math. Soc., (1972).

Mortensen, R.E. J. Statist. Phys. 1, 271-296, (1969).

Simpson-Herren, L. & Lloyd, H.H. Cancer Chemother. Rep. Part 1 54, 143-174, (1970).

Simpson-Herren, L. Personal communication, (1973).

Smith, C.E. Thesis, M.I.T., (1972).

Smith, C.E. & Tuckwell, H.C. J. Theor. Biol. (submitted), (1973)

Tuckwell, H.C. Theor. Pop. Biol. 5, (to appear), (1974).

Wong, E. & Zakai, M. Ann. Math. Stat. 36, 1560-1564, (1965).

DERIVATION OF THE EQUATION FOR CONCENTRATION PROFILES IN A BINARY DIFFUSING SYSTEM

George W. Swan
Department of Pure and Applied Mathematics
Washington State University
Pullman, Washington 99163, U.S.A.

ABSTRACT: Historically it appears to be a remarkable accident of chance that incorrect methods gave a correct equation for concentration profiles in a binary diffusing system. In this paper we highlight some of the logical inconsistencies of these methods and present a correct method which is free from error.

In the second part of the paper an application of this equation is presented to a mathematical model concerned with the transmission of information by the synapse.

Consider a volume of a mixture of fluids X and Y . Let

ρ_X = mass concentration of X (g of X/cm^3 of solution),

ρ_Y = mass concentration of Y (g of Y/cm^3 of solution),

$\underset{\sim}{v}$ = mass average velocity of the mixture.

The velocity vector has components (u,v,w) . We consider the case of a binary mixture which has a constant density and direct attention to the diffusion equation of a fluid flow of X in a host fluid Y . The rate of flow (or rate of transport) α of a substance with concentration $C(\underset{\sim}{r},t)$ ($\equiv \rho_X$, say) per unit area across any plane is given by Fick's first law of binary diffusion

$$\alpha = -D \frac{\partial C}{\partial n} , \tag{1}$$

where D is the molecular diffusion coefficient, and $\partial/\partial n$ denotes differentiation along the normal to the plane. The negative sign indicates that the transport occurs in the direction of decreasing concentration. If the concentration is the same everywhere no transport can take place. Thus, it is natural to consider in the first approximation, that the rate of transport of the dissolved substance at any point in a given direction is proportional to the rate of variation of the concentration at that point in the same direction.

The coordinate system $\underset{\sim}{r}$ is in a laboratory reference frame, that is we have current or Eulerian coordinates. The quantity D may be a constant or it may be a function of C or it may be a function of C and the first derivatives of C or it may depend on the time or spatial coordinates alone.

We consider a two dimensional configuration with diffusion in a rectangular shape (see Fig. 1) which is fixed in space. It is assumed that the substance within the rectangle is isotropic.

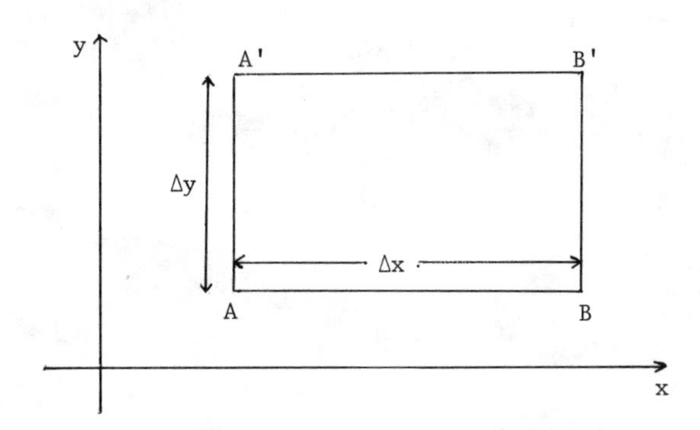

Figure 1: Diffusion in a rectangle

Here $AB = A'B' = \Delta x$, $AA' = BB' = \Delta y$, A is an arbitrary point (x_o,y_o) . Also, we assume that the mass within the element $ABB'A'$ is varying because of flow in and out across the walls; the fluid is assumed to have x- and y-velocity components given by the vector $\underset{\sim}{v} = (u,v)$.

In the following description it is assumed that all integrands are continuous functions of the independent variables throughout the domain of the multiple integrations. The amount of dissolved substance being transported across AA' into the rectangle in the time t_o to $t_o + \Delta t$ is given by

$$-\int_{t_o}^{t_o+\Delta t} \int_{y_o}^{y_o+\Delta y} \left(D \frac{\partial C}{\partial x}\right)_{x_o,y,t} dy\, dt \quad ,$$

where the suffix notation means that the quantity in parenthesis is to be evaluated at $x = x_o$, $y = y$, $t = t$. Also, the amount being transported across BB' out of the rectangle in the same interval of time is

$$-\int_{t_o}^{t_o+\Delta t} \int_{y_o}^{y_o+\Delta y} \left(D \frac{\partial C}{\partial x}\right)_{x_o+\Delta x,y,t} dy\, dt \quad .$$

Hence, the difference between the amount entering and the amount leaving is

$$\int_{t_o}^{t_o+\Delta t} \int_{y_o}^{y_o+\Delta y} F(\xi,y,t)\, dy\, dt \quad , \tag{2}$$

where

$$F(\xi,y,t) = \left(D \frac{\partial C}{\partial x}\right)_{x_o+\Delta x,y,t} - \left(D \frac{\partial C}{\partial x}\right)_{x_o,y,t}$$

$$= \left(\frac{\partial}{\partial x} D \frac{\partial C}{\partial x}\right)_{\xi,y,t} \Delta x \quad . \tag{3}$$

Also by assumption the quantity F is continuous in the domain of integration and we have used the derivative mean value theorem of calculus; $x_o < \xi < x_o + \Delta x$. In the same way, the difference between the amount entering across AB and the amount leaving across A'B' in the same interval of time is

$$\int_{t_o}^{t_o+\Delta t} \int_{x_o}^{x_o+\Delta x} G(x,\eta,t)\, dx\, dt \quad , \tag{4}$$

where

$$G(x,\eta,t) = \left(D \frac{\partial C}{\partial y}\right)_{x,y_o+\Delta y,t} - \left(D \frac{\partial C}{\partial y}\right)_{x,y_o,t}$$

$$= \left(\frac{\partial}{\partial y} D \frac{\partial C}{\partial y}\right)_{x,\eta,t} \Delta y \quad , \tag{5}$$

and again we have used the derivative mean value theorem; $y_o < \eta < y_o + \Delta y$.

If the dissolved substance is being produced in the solution due to chemical reactions, at a rate Q, then the total amount of its production inside the rectangle during the interval of time Δt is

$$\int_{t_o}^{t_o+\Delta t} \int_{x_o}^{x_o+\Delta x} \int_{y_o}^{y_o+\Delta y} Q \, dx \, dy \, dt \quad . \tag{6}$$

The total change of the amount of substance inside the rectangle is now the combination of (2), (4), and (6):

$$\int_{t_o}^{t_o+\Delta t} \int_{y_o}^{y_o+\Delta y} F(\xi,y,t) \, dy \, dt + \int_{t_o}^{t_o+\Delta t} \int_{x_o}^{x_o+\Delta x} G(x,\eta,t) \, dx \, dt$$

$$+ \int_{t_o}^{t_o+\Delta t} \int_{x_o}^{x_o+\Delta x} \int_{y_o}^{y_o+\Delta y} Q \, dx \, dy \, dt \quad . \tag{7}$$

The net contribution due to convection across BB' and AA' is

$$\int_{t_o}^{t_o+\Delta t} \int_{y_o}^{y_o+\Delta y} H(\xi_1,y,t) \, dy \, dt \quad , \tag{8}$$

where

$$H(\xi_1,y,t) = (uC)_{x_o+\Delta x,y,t} - (uC)_{x_o,y,t}$$

$$= [\frac{\partial}{\partial x}(uC)]_{\xi_1,y,t} \Delta x \quad . \tag{9}$$

Similarly, the net contribution due to convection across AB and A'B' is

$$\int_{t_o}^{t_o+\Delta t} \int_{x_o}^{x_o+\Delta x} I(x,\eta_1,t) \, dx \, dt \quad , \tag{10}$$

where

$$I(x,\eta_1,t) = (vC)_{x,y_o+\Delta y,t} - (uC)_{x,y_o,t}$$

$$= [\frac{\partial}{\partial y}(vC)]_{x,\eta_1,t} \Delta y \quad . \tag{11}$$

It follows that the net contribution due to convection is given by the sum of (8) and (9) as

$$\int_{t_o}^{t_o+\Delta t} \int_{y_o}^{y_o+\Delta y} H(\xi_1,y,t)\,dy\,dt + \int_{t_o}^{t_o+\Delta t} \int_{x_o}^{x_o+\Delta x} I(x,\eta_1,t)\,dx\,dt \quad . \tag{12}$$

The total change with time of the concentration is

$$\int_{t_o}^{t_o+\Delta t} \int_{x_o}^{x_o+\Delta x} \int_{y_o}^{y_o+\Delta y} \frac{\partial C}{\partial t}\,dx\,dy\,dt \quad . \tag{13}$$

On combining (7), (12) with (13) we obtain the balance law

$$\int_{t_o}^{t_o+\Delta t} \int_{x_o}^{x_o+\Delta x} \int_{y_o}^{y_o+\Delta y} \frac{\partial C}{\partial t}\,dx\,dy\,dt$$

$$+ \int_{t_o}^{t_o+\Delta t} \int_{y_o}^{y_o+\Delta y} H(\xi_1,y,t)\,dy\,dt + \int_{t_o}^{t_o+\Delta t} \int_{x_o}^{x_o+\Delta x} I(x,\eta_1,t)\,dx\,dt$$

$$= \int_{t_o}^{t_o+\Delta t} \int_{y_o}^{y_o+\Delta y} F(\xi,y,t)\,dy\,dt + \int_{t_o}^{t_o+\Delta t} \int_{x_o}^{x_o+\Delta x} G(x,\eta,t)\,dx\,dt$$

$$+ \int_{t_o}^{t_o+\Delta t} \int_{x_o}^{x_o+\Delta x} \int_{y_o}^{y_o+\Delta y} Q\,dx\,dy\,dt \quad ,$$

which becomes, on using the integral mean value theorem,

$$[\frac{\partial C}{\partial t} + H(\xi_1,y,t) + I(x,\eta_1,t) - F(\xi,y,t) - G(x,\eta,t) - Q]_{x_1,y_1,t_1} \Delta x\, \Delta y\, \Delta t = 0 \quad ,$$

where $x_o < x_1 < x_o + \Delta x$, $y_o < y_1 < y_o + \Delta y$, $t_o < t_1 < t_o + \Delta t$. The restriction on using this theorem is that the quantity within the square parenthesis must be continuous. Now cancel the $\Delta x\, \Delta y\, \Delta t$ and hence, in the limit as Δx , Δy , and Δt each tend to zero we obtain

$$\frac{\partial C}{\partial t} + \frac{\partial}{\partial x}(uC) + \frac{\partial}{\partial y}(vC) = \frac{\partial}{\partial x} D \frac{\partial C}{\partial x} + \frac{\partial}{\partial y} D \frac{\partial C}{\partial y} + Q \quad , \tag{14}$$

which holds at the point x_o,y_o . But this point was chosen arbitrarily, hence (14) holds at any point in the medium.

The quantity Q may be positive or negative, the latter case occurring when the dissolved substance is being removed (or being consumed). We talk of a source when Q is positive and a sink when Q is negative. Also Q may be zero, or a constant, or a function of the spatial coordinates, or a function of the concentration C .

The derivation of (14) is for a two dimensional region in space. In three dimensions for a velocity vector $\underset{\sim}{v} = (u,v,w)$ the corresponding equation for an isotropic medium is

$$\frac{\partial C}{\partial t} + \frac{\partial}{\partial x}(uC) + \frac{\partial}{\partial y}(vC) + \frac{\partial}{\partial z}(wC) = \frac{\partial}{\partial x} D \frac{\partial C}{\partial x} + \frac{\partial}{\partial y} D \frac{\partial C}{\partial y} + \frac{\partial}{\partial z} D \frac{\partial C}{\partial z} + Q \quad , \tag{15}$$

and the coordinate system is rectangular Cartesian. It is straightforward to convert (15) into the vector form

$$\frac{\partial C}{\partial t} + (\underset{\sim}{v} \cdot \mathrm{grad})C + C\, \mathrm{div}\, \underset{\sim}{v} = \mathrm{div}\, D\, \mathrm{grad}\, C + Q \quad . \tag{16}$$

This last form is especially convenient if we require to work in a coordinate system like cylindrical or spherical polars. An alternative form of (16) is

$$\frac{\partial C}{\partial t} + \mathrm{div}[C\underset{\sim}{v} - D\, \mathrm{grad}\, C] = Q \quad . \tag{17}$$

The reasons for the above derivation of (16) are as follows. Say we consider a typical way in which the continuity equation is produced for two dimensional unsteady flow.

There is the assumption that the mass flux into ABB'A' across AA' is given by $(\rho u)_A \Delta y$. (The difficulty here is that there is no guarantee that we are entitled to ignore the variation in ρu from A to A' .) With the same assumption, the mass flux out of ABB'A' across BB' is

$$(\rho u)_B \Delta y = \{(\rho u)_A + [\frac{\partial}{\partial x}(\rho u)]_A \Delta x + \cdots\}\Delta y$$

by a two term Taylor series expansion. In order to produce the quantities in the curved parentheses we are now allowing a variation in ρu from A to B , in direct conflict of the first assumption above! This type of derivation must therefore be regarded as unsatisfactory. The interesting thing is that the literature is full of such curious derivations and it is hoped that the derivation of this section, although lengthy, removes any objections.

We now present a biological example of the equation (16). The morphophysiological structure that provides the connection between two neurons is the synapse. It is believed that within the synapse information is carried by a chemical mediator, which is liberated by the presynaptic membrane. Consider a one dimensional situation with $C(x,t)$ denoting the mediator concentration. The time rate of change of the concentration is balanced by diffusion and destruction of itself. In the absence of convection, and with a constant coefficient of diffusion, (16) takes the form $C_t = DC_{xx} + Q$. If we express the removal of the mediator by a sink term with $Q = -bC$, b a positive constant, then

$$C_t = DC_{xx} - bC \quad , \quad t > 0 \quad , \quad 0 < x < 1 \quad , \tag{18}$$

where we take the boundaries as specified by $x = 0$, $x = 1$. Assume that, initially, the concentration of active substance within the synapse is zero. Also we assume that there is no flux across $x = 1$ whereas the flux across $x = 0$ is some function of the time. Thus,

$$C(x,0) = 0 \quad , \quad C_x(0,t) = f(t) \quad , \quad C_x(1,t) = 0 \quad . \tag{19}$$

The above problem, in a slightly different notation, is presented by Badescu (et al.), but their solution is not correct. Walsh (et al.) purport to state the correct solution, but it turns out that theirs is also incorrect. We proceed to give the details of the correct solution to this problem by a method different from that of Badescu et al.

Remove the inhomogeneous boundary condition on $x = 0$ by means of a function $v \equiv v(x,t)$ defined as

$$v(x,t) = k - C(x,t) - \frac{1}{2}(1-x)^2 f(t) \quad , \tag{20}$$

where k is some constant. The problem posed by (18), (19) now becomes

$$v_t + bv - Dv_{xx} = bk - \frac{1}{2}(1-x)^2[f'(t) + bf(t)] + Df(t) \quad , \tag{21}$$

with

$$v(x,0) = k - \frac{1}{2}(1-x)^2 f(0) \quad , \quad v_x(0,t) = v_x(1,t) = 0 \quad .$$

Because of the homogeneous conditions on v_x we introduce the finite cosine transform

$$A \equiv A(n,t) = \int_0^1 v(x,t)\cos n\pi x \, dx \quad . \tag{22}$$

Multiply (21) through by $\cos n\pi x$ and integrate the result with respect to x from 0 to 1. This gives

$$\frac{dA}{dt} + (b + Dn^2\pi^2)A = \frac{-1}{n^2\pi^2}[f'(t) + bf(t)] \quad , \quad n \neq 0 \quad .$$

Since $A(n,0) = -f(0)/n^2\pi^2$ we see that this equation has the solution

$$A(n,t) = \frac{-1}{n^2\pi^2} f(t) + De^{-(b+Dn^2\pi^2)t} \int_0^t f(\tau)e^{(b+Dn^2\pi^2)\tau} d\tau ,$$

which is true for $n \neq 0$. The equation for $A(0,t)$ is formed as follows. Multiply (21) through by $\cos 0 \equiv 1$ and integrate the result with respect to x from 0 to 1:

$$\frac{dA(0,t)}{dt} + bA(0,t) = bk - \frac{1}{6}[f'(t) + bf(t)] + Df(t) .$$

Since $A(0,0) = k - \frac{1}{6}f(0)$ this equation has the solution

$$A(0,t) = \{1 + (1-b)e^{-bt}\}k - \frac{f(t)}{6} + De^{-bt} \int_0^t f(\tau)e^{-b\tau} d\tau .$$

Inversion of (22) gives

$$v = A(0,t) + 2 \sum_{n=1}^{\infty} A(n,t)\cos n\pi x .$$

Hence we readily find that

$$v = \{1 + (1-b)e^{-bt}\}k - \frac{1}{2}(1-x)^2 f(t) + De^{-bt} \int_0^t f(\tau)e^{-b\tau} d\tau$$

$$+ 2D \sum_{n=1}^{\infty} \int_0^t f(\tau)e^{-(b+Dn^2\pi^2)(t-\tau)} d\tau \cos n\pi x ,$$

where we have made use of the result

$$\frac{1}{2}(1-x)^2 = \frac{1}{6} + \frac{2}{\pi^2} \sum_{n=1}^{\infty} \frac{\cos n\pi x}{n^2} , \quad 0 < x < 1 .$$

An expression for the mediator concentration is now obtained from (20) and we denote this expression by $\psi(x,t,k,b)$. Now the problem

$$\varphi_t = D\varphi_{xx} \quad , \quad \varphi(x,0) = 0 \quad , \quad \varphi_x(0,t) = f(t) \quad , \quad \varphi_x(1,t) = 0$$

has the solution

$$\varphi(x,t) = -D \int_o^t f(\tau)d\tau - 2D \int_o^t f(\tau) \sum_{n=1}^{\infty} e^{-Dn^2\pi^2(t - \tau)} d\tau \cos n\pi x \quad .$$

Since $\lim_{b \to o} \psi(x,t,k,b) = \varphi(x,t)$ we deduce that k must be identically zero and the final expression for the mediator concentration $C \equiv \psi(x,t,0,b)$ is

$$C(x,t) = -D \int_o^t f(\tau)e^{-b(t-\tau)}d\tau - 2D \int_o^t f(\tau) \sum_{n=1}^{\infty} e^{-(b+Dn^2\pi^2)(t-\tau)} d\tau \cos n\pi x \quad .$$

REFERENCES

Badescu, R., Balaceanu, C., and Nicolau, Ed., Bio-Medical Computing. 1, 211-220 (1970).

Walsh, R. A., and Prelewicz, D. A., Bio-Medical Computing. 2, 321-322 (1971).

AN ECONOMIC MODEL OF POPULATION GROWTH AND COMPETITION IN NATURAL COMMUNITIES

James E. Turner
Department of Mathematics
McGill University
Montréal, Québec

David J. Rapport
Department of Biological Sciences
Simon Fraser University
Burnaby, British Columbia

1. Introduction

A fundamental problem of ecology is to relate the dynamics of the growth and competition of biological populations to the underlying foraging process and to the detailed structure of the environment. The logistic growth model

$$\frac{dn}{dt} = rn(1 - \frac{n}{K}) \tag{1}$$

is the best known descriptive model of population growth in a limited environment. It describes a population coming into balance with its resources at the carrying capacity K of the environment. The Lotka-Volterra competition model (Slobodkin, 1961)

$$\begin{aligned} \frac{dn_1}{dt} &= r_1 n_1 (1 - \frac{n_1}{K_1} - \frac{an_2}{K_1}) \\ \frac{dn_2}{dt} &= r_2 n_2 (1 - \frac{bn_1}{K_2} - \frac{n_2}{K_2}) \end{aligned} \tag{2}$$

generalizes the logistic growth model to describe competition of two (or more) populations for limited resources. These models can be derived from varying assumptions concerning the dependence of birth and death rates on population densities (Pielou, 1969).

MacArthur (1972) has described the four essential ingredients of all interesting biogeographic patterns as the structure of the environment, the morphology of the species, the economics of species behavior, and the dynamics of population changes. In this paper, we model the dynamics of populations in terms of the economics of species behavior. A general substructure for growth and competition models is developed in terms of the energetics of foraging. Within this framework, the parameters of logistic growth and Lotka-Volterra competition are related to measurable energy variables describing relative efficiencies of resource utilization.

2. The Energetics of Resource Consumption

Efficiency in obtaining food resources is obviously an important factor in population growth and competition. In this section, we develop a model of a population growing on a single food resource available in limited but renewable supply. (For the purposes of this model, multiple food resources consumed in fixed proportions can be considered as a single food resource. The one resource model may be valid more generally when only one food resource is limiting.)

For an average predator, define C(q,n) to be the energy expenditure per unit of consumption when the consumption rate is q and the predator population is n. This energy cost includes capture, digestion, and replacement costs. Replacement cost is defined as the energy required for sufficient reproduction to maintain the predator population unchanged. Define $G(q,n)$ to be the corresponding energy gain per unit of consumption. Then $\Pi(q,n) = q[G(q,n) - C(q,n)]$ is the total energy surplus (or deficit) for an average predator above maintenance requirements. We interpret $\Pi(q,n)$ as the energy available for net reproduction to an average predator.

To introduce a possible dynamics of the predator population, we assume that the average growth rate of the predator population is a function of the energy surplus to maintenance requirements of an average predator.

$$\frac{1}{n}\frac{dn}{dt} = f(\Pi(q,n)) \tag{3}$$

From the biological interpretation of this model, it is clear that f is a monotone increasing function of its argument satisfying $f(0) = 0$. To have an explicit model, we assume that the average growth rate is proportional to the energy surplus available for net reproduction.

$$\frac{1}{n}\frac{dn}{dt} = \gamma\Pi(q,n) \tag{4}$$

The parameter γ is a positive constant which measures the efficiency of converting an energy surplus to population growth.

To illustrate the applications of this general model of population growth, we assume reasonable simple forms for the functions $C(q,n)$ and $G(q,n)$. In particular, we assume that $G(q,n)$, the assimilated energy per unit of consumption, is a constant ρ. This is a reasonable assumption near equilibrium or over a limited range of consumption rates and population sizes. Further, we assume that

$$C(q,n) = \alpha(q - \mu)^2 + \beta n + \varepsilon \tag{5}$$

where α, β, ε, and μ are positive constants. In this model, μ is the consumption rate of minimum cost per unit of consumption, $\beta = \frac{\partial}{\partial n} C(q,n)$ measures the strength of intraspecies competition, and ε represents a threshold cost of consumption. As q increases above μ, $C(q,n)$ increases due to inefficiencies at high consumption rates caused by resource scarcity or digestion limitations or similar factors.

The optimal consumption rate $q_\circ(n)$ when the predator population is n, is defined as the consumption rate that maximizes the energy surplus available for reproduction. It is determined by

$$\frac{\partial}{\partial q}\Pi(q,n) = \frac{\partial}{\partial q}[q[G(q,n) - C(q,n)]] = 0 \tag{6}$$

The dynamics of optimal growth are given by

$$\frac{dn}{dt} = \gamma n\Pi(q_\circ(n),n) = \gamma n q_\circ(n)[G(q_\circ(n),n) - C(q_\circ(n),n)] \tag{7}$$

In the model of equation *(5)*, the optimal consumption rate is

$$q_\circ(n) = \frac{2}{3}\mu + \frac{1}{3}\sqrt{\mu^2 + 3\frac{(\rho-\varepsilon-\beta n)}{\alpha}} \tag{8}$$

Equation *(7)* for this explicit model is a complicated first order equation describing the growth of the predator population to the equilibrium $\hat{n} = \frac{\rho-\varepsilon}{\beta}$.

The consumption behavior of the predator population in response to the availability of resources will generally be more complex than that assumed in the optimal growth model. Populations which respond too quickly to increased abundance of resources may be selected against in variable environments. An alternative growth strategy would be to consume at the rate of minimum average cost per unit of consumption. In the explicit model, this generates the dynamics

$$\frac{dn}{dt} = \gamma\mu n(\rho - \varepsilon - \beta n) \tag{9}$$

This is the logistic growth model of equation *(1)* if we identify the constants $K = \frac{\rho - \varepsilon}{\beta}$ and $r = \gamma\mu(\rho - \varepsilon)$.

If $|\frac{\rho - \varepsilon - \beta n}{\alpha\mu^2}| << 1$, equation *(7)* can be expanded in powers of $\frac{\rho - \varepsilon - \beta n}{\alpha\mu^2}$. Ignoring second order terms, we obtain

$$\frac{dn}{dt} = \gamma\mu n(\rho - \varepsilon - \beta n)\ (1 + \frac{\rho - \varepsilon - \beta n}{4\alpha\mu^2}) \tag{10}$$

The factor $(1 + \frac{\rho - \varepsilon - \beta n}{4\alpha\mu^2})$ is a correction term obtained by modifying the consumption rate from μ to the optimal rate $q_o(n)$. The effect of this modification is to produce a more rapid approach to equilibrium than in the logistic growth model.

3. Competition for Resources

The energetics framework of resource consumption developed in the previous section can be extended to model competition for one or more resources. For simplicity, consider two populations n_1 and n_2 competing for a single food resource. Define q_1 and q_2 to be the consumption rates of average predators of the two populations. Generalizing the model of the previous section, we assume that the energy costs per unit of consumption are given by

$$\begin{aligned} C_1(q_1,q_2,n_1,n_2) &= \alpha_1(q_1 - \mu_1)^2 + \beta_{11}n_1 + \beta_{12}n_2 + \varepsilon_1 \\ C_2(q_1,q_2,n_1,n_2) &= \alpha_2(q_2 - \mu_2)^2 + \beta_{21}n_1 + \beta_{22}n_2 + \varepsilon_2 \end{aligned} \tag{11}$$

The parameters of this model are positive constants. If both populations adopt the strategy of consuming at the rate of minimum average cost, then $q_1 = \mu_1$ and $q_2 = \mu_2$. At these consumption rates, the energy surpluses for average predators are

$$\begin{aligned} \Pi_1(\mu_1,\mu_2,n_1,n_2) &= \mu_1[\rho_1 - \varepsilon_1 - \beta_{11}n_1 - \beta_{12}n_2] \\ \Pi_2(\mu_1,\mu_2,n_1,n_2) &= \mu_2[\rho_2 - \varepsilon_2 - \beta_{21}n_1 - \beta_{22}n_2] \end{aligned} \tag{12}$$

Introducing the dynamics of the previous section, the growth rates of the competing populations are given by

$$\begin{aligned} \frac{dn_1}{dt} &= \gamma_1\mu_1 n_1[\rho_1 - \varepsilon_1 - \beta_{11}n_1 - \beta_{12}n_2] \\ \frac{dn_2}{dt} &= \gamma_2\mu_2 n_2[\rho_2 - \varepsilon_2 - \beta_{21}n_1 - \beta_{22}n_2] \end{aligned} \tag{13}$$

This is the Lotka-Volterra model of equation *(2)* if we identify the constants of the two models.

$$r_1 = \gamma_1\mu_1(\rho_1 - \varepsilon_1) \quad K_1 = \frac{\rho_1 - \varepsilon_1}{\beta_{11}} \quad a = \frac{\beta_{12}}{\beta_{11}}$$

$$r_2 = \gamma_2\mu_2(\rho_2 - \varepsilon_2) \quad K_2 = \frac{\rho_2 - \varepsilon_2}{\beta_{22}} \quad b = \frac{\beta_{21}}{\beta_{22}} \tag{14}$$

We are now able to interpret the parameters of Lotka-Volterra competition. The biotic potential r_1 is proportional to the consumption rate μ_1 and to the difference $\rho_1 - \varepsilon_1$ between the energy return and the threshold energy cost of consumption of a unit of the resource. The parameters $\beta_{ij} = \frac{\partial C_i}{\partial n_j}$ measure the relative importance of intraspecies and interspecies competition for resources. The carrying capacity K_1 is proportional to $\rho_1 - \varepsilon_1$ and inversely proportional to β_{11} which measures the intensity of intraspecies competition in the first population. The interaction coefficient a is the ratio of two energy cost coefficients, β_{12} and β_{11}. This ratio measures the relative costs to the first population of increases in the second and first populations.

The four possible outcomes correspond to the following inequalities.

$$\frac{\beta_{12}}{\beta_{22}} > \frac{\rho_1 - \varepsilon_1}{\rho_2 - \varepsilon_2} > \frac{\beta_{11}}{\beta_{21}} \qquad \text{Indeterminate outcome}$$

$$\frac{\beta_{12}}{\beta_{22}} < \frac{\rho_1 - \varepsilon_1}{\rho_2 - \varepsilon_2} < \frac{\beta_{11}}{\beta_{21}} \qquad \text{Coexistence}$$

$$\frac{\beta_{12}}{\beta_{22}}, \frac{\beta_{11}}{\beta_{21}} < \frac{\rho_1 - \varepsilon_1}{\rho_2 - \varepsilon_2} \qquad \text{First species wins}$$

$$\frac{\beta_{12}}{\beta_{22}}, \frac{\beta_{11}}{\beta_{21}} > \frac{\rho_1 - \varepsilon_1}{\rho_2 - \varepsilon_2} \qquad \text{Second species wins}$$

We observe that the three ratios $\frac{\beta_{12}}{\beta_{22}}$, $\frac{\beta_{11}}{\beta_{21}}$ and $\frac{\rho_1 - \varepsilon_1}{\rho_2 - \varepsilon_2}$ determine the outcome of competition. Coexistence occurs if intraspecies competition dominates interspecies competition (β_{11} and β_{22} large compared to β_{12} and β_{21}). The outcome of competition depends critically on the relative energy returns, ρ_1 and ρ_2. If these assimilated energies per unit consumption are variable because of seasonal or other factors, the outcome of competition is also variable.

4. Discussion

We have developed a general substructure for dynamical models of population growth and competition in terms of the energetics of resource consumption. This framework allows the logistic growth model and the Lotka-Volterra competition model to be related to the underlying processes of foraging for a limited resource. A number of testable conclusions have been generated concerning the dependence of the biotic potential and the carrying capacity on parameters related to the energetics of the foraging process.

The model may be extended to take into account more than one food resource and other limiting factors such as space. More complex strategies of resource consumption can be modelled by introducing concepts from control theory and differential games. For example, a possible competitive strategy would be to consume at rates above the optimal rate in order to deny the limited resource to competitors.

ACKNOWLEDGEMENTS

This work was supported by a Canada Council I. W. Killam Senior Research Scholarship.

REFERENCES

MacArthur, R. H. Geographical Ecology. Harper and Row, New York (1972).

Pielou, E. C. An Introduction to Mathematical Ecology. Wiley-Interscience, New York (1969).

Slobodkin, L. B. Growth and Regulation of Animal Populations. Holt, Rinehart and Winston, New York (1961).

WAVELIKE ACTIVITY IN BIOLOGICAL AND CHEMICAL MEDIA

Arthur T. Winfree
Department of Biological Sciences
Purdue University
West Lafayette, Indiana 47907

INTRODUCTION

A variety of room-temperature aqueous solutions colorfully exhibit in two and three dimensions the excitability, signal transmission, and spontaneous oscillation more familiarly associated with biological media such as aggregating slime molds, rhythmic fungi, nerve, heart, and intestinal smooth muscle. I will refer to these chemical solutions collectively as "Z reagent" in honor of their chief developers, A. M. Zhabotinsky and A. N. Zaikin (1970). Because Z reagent is so convenient to prepare and study, it is tempting to employ it as a heuristic tool in seeking consequences of oscillation and excitability which may have escaped attention during technically more difficult experiments in living systems. My belief is that observations on Z reagent suggest at least one unexplored mode of solution to the partial differential equations used to describe excitable media, which may merit the attentions both of mathematicians and of biologists.

Accordingly, I take this opportunity for a brief survey of Z reagent behavior, emphasizing analogies to biological media. I will try to cite as much recent literature as feasible since the pertinent journals and symposia are usually found in separate rooms of the library.

In a nutshell, Z reagent is a colored solution which develops moving red and blue bands. For accounts of the chemical basis of this spatially-patterned reaction see Field, Köros, and Noyes (1972); Field and Noyes (1972); Field (1972). In studying the moving patterns, I find it useful to distinguish "pseudo-waves" from "trigger waves" (or "waves"), and to distinguish two kinds of trigger-wave source: "target" nuclei, and a rotating dissipative structure that I will call a "scroll core" or when appropriate, a "scroll ring". In making these distinctions I am struggling to clarify in my own mind a point of particular confusion (as I see it) in the current literature: that several theoretical models of Z waves have been stimulated by observations on trigger waves, but in fact describe pseudo-waves.

PSEUDO-WAVES

Several models have appeared in the literature surrounding Z reagent, which thoroughly develop the properties of what I (1972) call pseudo-waves, Kopell and Howard (1973) call kinematic waves, Ortoleva and Ross (1973) call phase waves, etc. These are the visible expression of shallow concentration gradients which result in gradient of timing of a limit cycle oscillation parochially pursued in each volume element of the medium ---- like the pseudo-wave of traffic-light switching that gates traffic along major avenues in a city. Pseudo-waves of nearly synchronous mitosis are familiar in diverse organisms e.g. Physarum (Cohen (1972)). Pattern polymorphism in a growing fungus has been described as a consequence of shallow phase gradients in a sheet of nearly independent limit-cycle oscillators (Winfree (1970), (1973)). Following the lead of Beck and Varadi (1972), Kopell and Howard (1973), Smoes and Dreitlein (1973) and Thoenes (1973) have modelled Z reagent in this way, additionally allowing phase gradients to increase due to a spatial gradient in some parameter (e.g. temperature) which governs the period of local oscillation. By an elegant experiment Kopell and Howard (1973) and Thoenes (1973) independently confirmed the claim of Beck and Varadi (1972) that the horizontal color bands which appear in a vertical column of Z reagent (Busse (1969)) in some cases constitute not a dissipative structure as in Herschkowitz-Kauffman (1970), Glansdorf and Prigogine (1971), but pseudo-waves organized by a vertical gradient of temperature or chemical concentration.

Ortoleva and Ross (1973) have constructed a more detailed theory of phase gradients modified by diffusion near a heterogeneity.

In Z reagent, pseudo-waves are characterized by four features:

1) Velocity depends on position and can be arbitrarily large; given $\phi = \phi(x,t)$, velocity $= \left.\frac{dx}{dt}\right|_{\phi} = -\partial\phi/\partial t \;/\; \partial\phi/\partial x = -2\pi/\tau\partial\phi/\partial x$ in the direction of the

phase gradient. This velocity exceeds all bounds wherever the medium oscillates homogeneously ($\partial\phi/\partial x = 0$); i.e. the bulk oscillation _is_ a very fast pseudo-wave and a pseudo-wave _is_ a gradient of bulk oscillation phase. The time between waves at a fixed station in the reagent is exactly τ, the fixed period of the local oscillation. A packet of wider-spaced waves travels proportionately faster: wavelength/velocity = τ. (If, as in Smoes and Dreitlein (1973), a period-governing parameter diffuses, the local τ may be changing slowly, so the above will be true only in approximation).

2) In a rotating wave with radial wavefront (e.g. near the center of a spiral), velocity normal to the wavefront is proportional to distance from the pivot (c.f. DeSimone, Beil, and Scriven (1973)).

3) Being entirely determined by local oscillations, a pseudo-wave is not blocked by impermeable barriers (Winfree (1972); Kopell and Howard (1973)).

4) The wavefront can be indistinct and mottled, particularly where velocity is large in nearly homogeneous regions.

SPIRAL PSEUDO-WAVES

In theory, pseudo-waves can assume almost any shape, including concentric rings ("target patterns", corresponding to a radial phase gradient) and any number of radial arms (spirals) pivoting around a point (corresponding to a circular phase gradient). Spiral pseudo-waves have not been reported in Z reagent, but I am confident that they will be. The discussion to follow in this section anticipates some of their properties by analogy to other media (mostly biological) and by reference to recent theoretical papers.

Unlike Thoenes (1973) I do not believe that a continuous spatial distribution of frequency suffices to generate spiral activity patterns in a medium initially in synchronous limit-cycle oscillation. The reason is that the implied circular phase gradient requires an interior phase_less_ point (Winfree (1973)), for example the oscillation's stationary-state. So long as there are no holes in the medium and every volume element oscillates with a well-defined period, no phaseless states can be realized, thus no circular phase gradients, and so no rotating pseudo-waves. For the same reason, regardless of how it may have been formed, it is not possible to describe a rotating (spiral or radial) wave as a spatial distribution of phases on a common limit cycle: volume elements near the pivot cannot be on the limit cycle.

However it is possible to describe the formation of rotating pseudo-waves by processes which lead some oscillators far from the usual cycle. Three such processes will be described briefly. What they have in common is that the chemical concentration along any closed curve encircling the wave's rotation axis must wind once or more around the equilibrium state in concentration space, i.e. have non-zero winding number. An implication is that the rotation axis is at equilibrium, or at least not oscillating on the same cycle.

1) Spiral pseudo-waves from an initially homogeneous equilibrium: Without invoking diffusion or parameter gradients such as a spatial distribution of periods, spiral pseudo-waves can be formed from nearly homogeneous initial conditions close to an oscillator's unstable equilibrium. It is only necessary that the image in concentration space of some closed curve in real space should have non-zero winding number, $n \neq 0$, about the equilibrium. Though concentrations everywhere start quite close to equilibrium, none cross over it; the equilibrium being a repellor by assumption. Thus the winding number cannot change while volume elements diverge from equilibrium to a limit cycle. The result is a circular phase gradient through n full cycles of phase; this is a rotating wave with n arms (see discussion of pattern polymorphism in the fungus Nectria, Winfree (1970), (1973)).

2) Spiral pseudo-waves by graded perturbation: A spiral pseudo-wave can also be formed from a segment of pseudo-wave by "breaking" it in a transient parameter-gradient parallel to the wavefront; it is not necessary to invoke diffusion. I take a biological example from (Winfree (1970b):the 'Pinwheel Experiment'): a square grid composed of tens of thousands of Drosophila pupae harboring circadian (24^h) oscillators can be phased so that a pseudo wave of emergence activity propagates West to East. Sketch 1 ((PLACE SPACE, BEFORE) to

(STATE SPACE, BEFORE)) shows the mapping of each pupa from its place in this grid to its state in the concentration space of the circadian oscillation. Then a North-South gradient of light is imposed, which moves the southmost oscillators toward the bottom of their state plane as in Sketch 1 ((STATE SPACE, BEFORE) to STATE SPACE, AFTER)). Thus the boundary of this grid of pupae has its winding

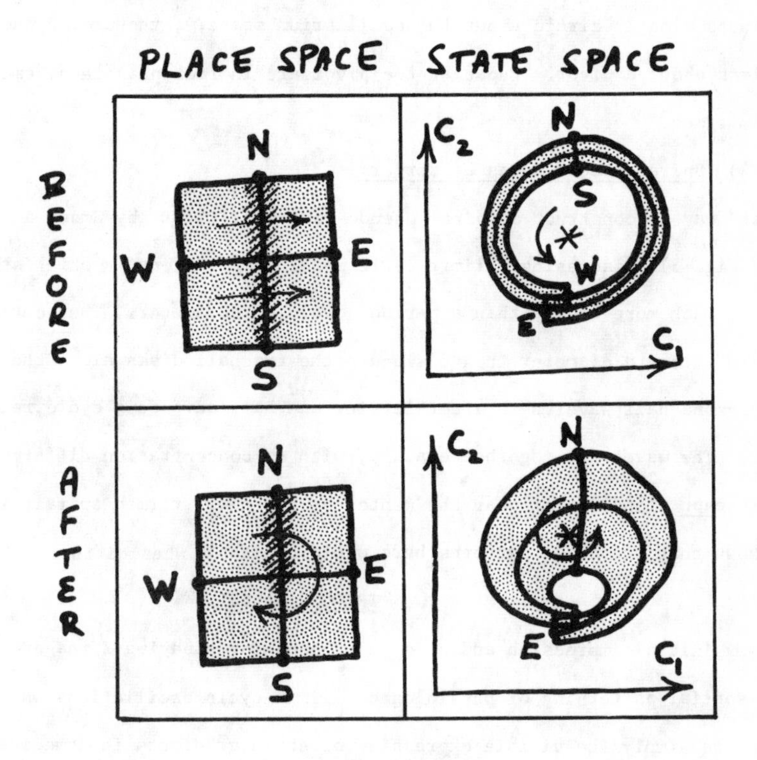

Sketch 1:

BEFORE	PLACE SPACE	A grid of pupae with circadian 'clock' phases 0 to 24 hours along an East-West gradient. A pseudo-wave of emergence sweeps from left to right in 24 hours.
	STATE SPACE	Each pupa is mapped to its instantaneous position (C_1,C_2) in the 'clock' oscillator's concentration space. Each independently follows a closed circular trajectory with 24^h period around the equilibrium state, *. (No limit cycle).
AFTER	PLACE SPACE	The South edge is nearly synchronized, the North edge unaffected, and intermediate pupae perturbed to intermediate extents. One pupa is necessarily left at the equilibrium, *.
	STATE SPACE	Looking again at each pupa in its place on the grid, we see that as all pupae continue to oscillate at 24^h period, emergence activity circulates about the pupa which was perturbed to equilibrium.

number about equilibrium changed from 0 to 1. Now suppose emergence occurs only when the circadian oscillator is at the top of its orbit, and that all orbits are concentric circles (which supposition is needlessly stringent, but convenient in present context). Emergence now occurs only in the top half of the North-South midline at this time and at 24 hour intervals after. As each pupal oscillator continues along a circle about the equilibrium state,*, the broken pseudo-wave revolves about a pivot. Pupae at the pivot are in fact phaseless, emerging at random times.

3) Spiral pseudo waves by surgery:

A third way to construct a spiral pseudo-wave was proposed by Smoes and Dreitlein (1973 Fig. 5), discussing a limit-cycle model of Z reagent in which state variables diffuse much more slowly than a period-governing parameter. They cut a "target pattern" along a diameter then dislocate the two half disks along the cut by exactly one half wavelength after letting one half advance ½ cycle relative to the other. The waves fit together exactly, with no concentration differences across the cut (*except* along a slit near the center of the newly formed spiral; it remains to be shown that the phase patterns here would be stable when diffusion is admitted).

TRIGGER WAVES

My initial impression and original reason for studying Z reagent in connection with spatial patterning of physiological limit-cycle oscillations was that *all* Z waves are simply the visible expression of phase gradients in a medium given to parochial oscillation, and that the role of diffusion is only to ensure continuity and gradually to level both parameter gradients and phase gradients. To prove this, I adjusted the chemical recipe to suppress spontaneous oscillation, and with it, all pseudo-waves*. But the most conspicuous waves persisted!

---

*The experimental "proof" offered in Winfree (1972) that the limit-cycle had been eliminated was that isolated droplets of reagent remained red for a long time, with the exception of a few that contained pacemaker nuclei. I have since noticed that oscillation is inhibited by contact with air in very thin layers, rendering this experiment inconclusive (See also Beck and Varadi (1972)). In fact, the reagent does oscillate in bulk with a period around 15 minutes .... still much longer than the time scale of concern, but not infinite; in Millipore stacks it oscillates with 3 to 5 minute period. However the same recipe but for a 50% decrease in sulfuric acid is definitely quiscent until stimulated, whereupon it propagates trigger waves normally.

These are the target-like circular waves described by Zaikin and Zhabotinsky (1970) and the spiral waves described by Winfree (1972) and Zhabotinsky and Zaikin (1973), and the scroll-shaped waves described by Winfree (1973b). Following Zeeman's discussion of wave propagation is excitable media (1972) I call them "trigger waves" or just "waves". In Z reagent, trigger waves are characterized by four features:

1) They occur in a medium which does not spontaneously oscillate, as well as in oscillating reagent; a <u>finite</u> disturbance is needed to trigger a wave.

2) Each leading edge propagates at a constant <u>finite</u> velocity determined by the state of the medium just ahead. The speed of a given wave, rather than the period of a given volume element, stays roughly constant.

3) The leading edge is always a remarkably sharp red/blue transition. It is not noticeably less sharp in waves 5 cm. apart than in waves 1 mm. apart.

4) The leading edge is instantly blocked by any barrier to molecular diffusion (Winfree (1972); DeSimone, Beil, and Scriven (1973) Fig. 2, but not theory) or a temporary chemical blockade (Winfree (1974)). The more gradual blue/red transition behind the leading edge is <u>not</u> blocked by a barrier quickly planted behind the leading edge. This suggests that only the leading edge really propagates: the return of red is dominated not by diffusion but by local kinetics (c.f. Zeeman (1972)).

It is my impression that most of the chemical literature cited above was stimulated by Zaikin and Zhabotinsky's 1970 report which used an oscillating reagent derived from Belousov (1958). Up to that time, I believe the only 'waves' reported were pseudo-waves in Belousov's reagent. And though the descriptions of trigger waves are cited, the theoretical models imputed to them by Thoenes (1973), and by Smoes and Dreitlein (1973), seem to me to describe pseudo-waves more accurately.

I find the current literature quite confusing in this respect, and believe that it <u>will</u> be necessary to distinguish between kinds of wave, whether or not in the way here suggested.

Trigger waves are common in excitable media of biological interest, action potentials being the most conspicuous example in nerve, heart, muscle, and some plants. The cells of social amoeba Dictyostelium discoideum also constitute an excitable but not spontaneously oscillatory medium: each cell responds to extracellular "acrasin" (believed to be cyclic AMP) by emitting a pulse of it. Thus an electrophoretic pulse of c-AMP at a micropipette tip triggers a circular wave which expands across the sheet of cells at 40 microns per minute (Robertson, Drage, and Cohen (1972)). Occasional isolated cells do emit acrasin pulses spontaneously at 5 minute intervals; they become sources of "target" wave patterns. Because cells migrate toward wave sources, each such pacemaker soon collects enough cells to form a slug, which accordingly crawls off to differentiate and reproduce (Robertson (1972)).

TRIGGER WAVE VELOCITY

It should be possible to calculate the propagation velocity from diffusion coefficients and the rate constants of the elementary processes composing a medium's excitability. For nerve it was done by Hodgkin and Huxley (1952). Cohen and Robertson (1971) have done this for Dictyostelium. Field and Noyes (1974b) have undertaken the calculation for Z reagent.

Plane waves are stable against local irregularities, for merely geometrical reasons best expressed in a drawing. Sketch 2 shows the evolution of a notched wave with fixed propagation velocity along the local wavefront normal: the notches are eliminated.

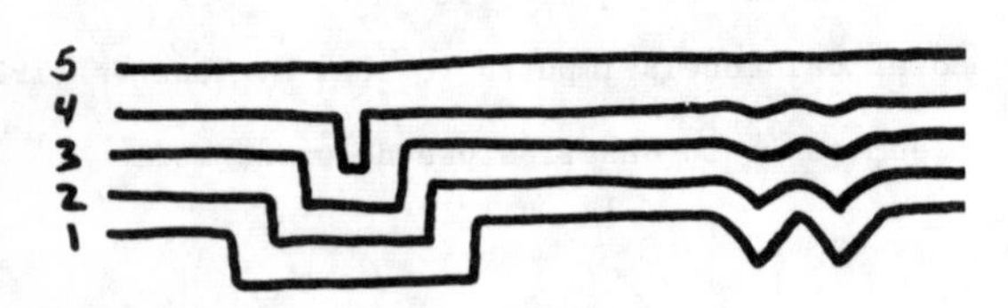

Sketch 2: The curves labled 1,2,3, ... etc. are successive positions of a plane wavefront. It is assumed that the wave propagates unit distance along the local normal in unit time. Thus irregularities are smoothed over.

However the spiral eigenpattern of linear diffusion-reaction kinetics suggests to DeSimone, Beil, and Scriven (1973) an additional dependence of normal velocity upon local wavefront curvature. This seems reasonable physically: the concentration of material diffusing from a curved source rises faster on the inside than on the outside of the curve. However, in Z reagent, I have been unable to measure such an effect at radii of curvature down to 1 mm.

Gulko and Petrov (1972) report changes of computer-simulated nerve wave velocity in local irregularities, but it is not clear whether they refer to the geometrical effect of Sketch 2, a physical effect like DeSimone's, or something else.

As a next project, it should be possible to calculate the dispersion relations: the velocity of a trigger wave travelling a given distance behind a forerunner. (Pseudo-wave velocity is simply a fixed constant divided by the distance between waves, since the temporal period is the same everywhere.) In Z reagent, trigger waves are noticeably slower when riding close upon their predecessors coat-tails, where the reagent has not fully relaxed back to the quiescent red state. Thus a target pattern dilates as it expands, the gap between the first two waves widening faster than more interior gaps.

I have argued elsewhere (Winfree (1974)) that Z reagent approximates an aqueous solution of the Hodgkin-Huxley equations for nerve excitability. Rinzel and Keller (1973) have calculated a dispersion relation resembling sketch 3 for a simpler caricature of those equations which retains the same qualitative behavior. If such a relation obtains in Z reagent and waves on arc AB are stable, then the description of spiral waves as the maximum frequency, minimum wavelength mode (Winfree (1972)) cannot be correct: for those are respectively the points (A) of radial tangency and (B) of 90° tangency to the velocity curve. Because they eventually dominate all other sources, it seems likely that spirals have maximum frequency (A) rather than minimum wavelength (B) on such a curve. Lower-frequency source in range $\sphericalangle$ AOB may have two stable velocities (on AB and AC), but only the faster one will be realized as the gaps ahead of successive waves follow the zigzag path from infinity at v_{max}.

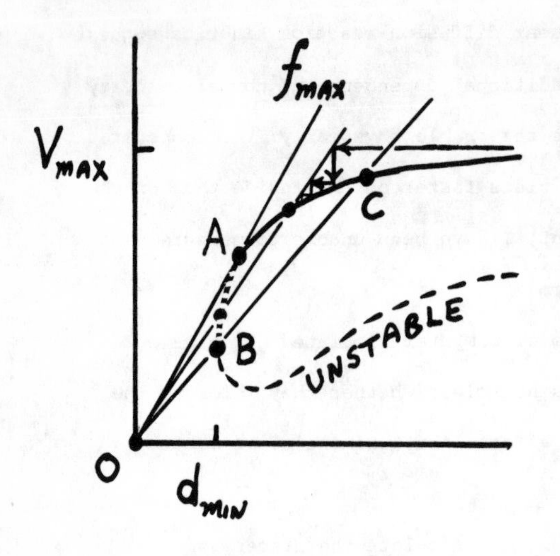

Sketch 3: A speculative anticipation of the relation between Z wave velocity and its distance behind a preceeding wave. Point A has maximum frequency, supposing a periodic wave train; Point B has maximum wave number. Tatterson and Hudson (1973) have measured a portion of this curve. In Rinzel's model, the lower (unstable) arc extends beyond B, possibly as far as A (Rinzel and Keller (1973)).

Trigger waves propagate slower in very thin films, but this is due to exposure to air: a plane wave propagates without deformation under a glass lens touching the bottom of the dish. This seems reasonable considering that interfaces serve essentially as "mirrors" to diffusion, as in electrostatic virtual imaging: because there is no transport across an interface, it serves the same function as a mirror-image slab of reagents. Thus thickness should make little difference for plane-wave propagation parallel to parallel interfaces (so long as there is no transport across them).

TRIGGER WAVE SOURCES

Trigger waves emerge from at least two kinds of source in Z reagent: from target centers, and from free edges of waves.

1) Target centers: Heterogeneous nuclei are often visible in the center of target patterns, sending out circular waves at an interval shorter than bulk oscillation (pseudo-wave period) and longer than scroll core rotation (see below). Sprinkling dust into Z reagent nucleates lots of new target patterns. Filtering reagent free of particles $> 1/5\ \mu$ diameter eliminates the majority of them.

Some still emerge from scratches in the container wall; lining the dish with silicone eliminates most of these*. However Zhabotinsky and Zaikin (1973) report quite different results in similar experiments, and suggest that target patterns may not require heterogeneous nuclei.

Any other condition that induces local excitation (blue color) of a sufficient volume will trigger a circular wave; 10^{-5} ml. is sufficient. The exposed filament of a pocket flashlight, touching the reagent surface or even separated from it by a glass coverslip, dispenses one circular wave each time the switch is momentarily closed (so long as a minimum interval, about one spiral wave rotation period (see below) is allowed between pulses). In non-oscillating reagent <u>only</u> one wave results from one stimulus. However after spontaneously <u>oscillating</u> reagent has once been stimulated at a point, each volume element of reagent begins its cycle to the next spontaneous bluing when the trigger wave goes through it. So the trigger wave is repeated as a pseudo-wave of identical shape and velocity, again and again at the period of the spontaneous bulk oscillation.

Herschkowitz-Kauffman and Nicolis (1972) have proposed a mechanism whereby a non-oscillating medium with appropriate diffusion coefficients might spontaneously oscillate only in a limited volume positioned by the container interfaces. If such a spot arose in scratches etc. in the container wall, it might serve as a target nucleus. Zhabotinsky and Zaikin (1973) discuss a similar model.

2) <u>Free edges of waves</u>: The other kind of source is not a point, but a line, which forms near any free edge of a trigger wave. Imagine a trigger wave expanding as a blue cylindrical shell through otherwise red (quiescent) Z reagent. If the reagent is gelled with colloidal SiO_2 or immobilized in Millipore cellulose matrix, a volume of red reagent can be layered on top without deforming the wave. The chemical concentration isobars constituting the cylindrical wave are forced to

---

* The abundance of target nuclei, or "pacemakers", depends on reagent excitability. The recipe of (Winfree (1972)) can be made less excitable by using ½ as much sulfuric acid; filtered into silicone-lined dish it lies completely inert until deliberately triggered, where upon a single wave propagates normally.

make a topological readjustment with the result that the free edge of the cylinder begins to curl up around a horizontal circular axis. Thus a scroll-shaped wave develops, whose axis is a ring. The scroll's rotation period is shorter than the pseudo-wave oscillation, and seems to be shorter than that of any pacemaker nucleus. Since waves emitted from two sources collide closer each time to the longer-period source, the scroll wave eventually displaces all other sources. All scroll waves have approximately the same period. The only sources surviving after some dozens of rotations are these stationary rings (many of them interruped by the container's walls, especially in thin layers of reagent).

A scroll ring looks a lot like a vortex ring of rotating fluid, in which a streamer of smoke or dye has been allowed to curl up into a spiral. However in the scroll ring, nothing moves physically; only a wave of activity circulates in a motionless fluid. In a vortex ring, the smoke sheet coils tighter and tighter because the inner laminae of the torus rotate faster than the outer ones; but in a scroll ring the spiral has fixed pitch and rotates rigidly.

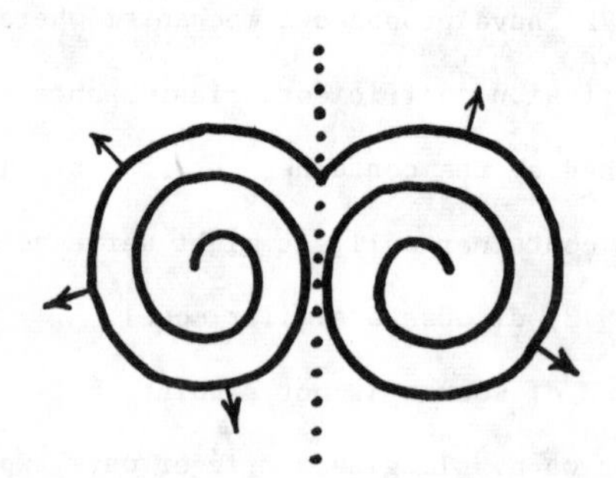

Sketch 4: The black curves represent wavefronts (abrupt red-to-blue transitions) emerging from two nearby counter-rotating spirals on a plane. Rotated about the (dotted) mirror axis, these curves become surfaces of revolution resembling waves emerging from a scroll ring.

A scroll ring dispenses waves which, beyond a certain distance, form concentric closed shells, as seen in cross-section in Sketch 4, above. In principle, it might seem that the ring could contract to a point and vanish, leaving an expanding halo of concentric spheres. However I've never seen this happen; even a ring about one wavelength in circumference seemed perfectly stable during the fifteen cycles I watched. However it can gradually drift into a container wall until, when the last arc of scroll axis vanishes, all new waves cease to appear. The existing nest of concentric hemispheres continues to propagate away, as a hole of quiesence expands inside the innermost. This hole is eventually occupied by waves from a nearby scroll, dilating somewhat as they propagate into unoccupied medium.

In spontaneously oscillating reagent, the phase gradient set up by the last wave from the scroll axis remains as a pseudo-wave, repeated at the oscillation frequency and forming a target pattern.

In theory, it might seem that scroll rings could be fashioned with the scroll given any integer number of full 360° twists before its axis is closed in a ring; or that the axis could be knotted. Such scroll rings would be topologically distinctly different in all cross-sections. But I have seen only the simplest kind: Sketch 4 made into a surface of revolution about the dashed axis (the "hole axis"). The scroll axis is seldom a perfect circle, but I have never knowingly encountered a knot or a twist.

Are the other species of scroll ring unstable? Or impossible for some topological reason? Or have I only failed to arrange suitable initial conditions? The question invites and awaits study.

SPIRAL TRIGGER WAVES IN LIVING SYSTEMS

So far as I am aware, three-dimensional scroll waves have not been observed in any excitable medium of biological interest. However in two dimensions aggregating cell fields of Dictyostelium discoideum, the "social amoeba", exhibit spirals which resemble scroll waves in plane section normal to the scroll axis. The analogy is closest in dense fields of a non-aggregating mutant (Gerisch(1965), (1968), (1971)). In other strains (e.g. Durston's NC-4 (1974)) the medium's properties are affected (1) by the passage of waves: cells migrate toward wave sources, resulting in density instabilities; and (2) by aging of cells: their refractory period diminishes , and with it, the spirals' rotation period. Dictyostelium additionally differs from Z reagent in supporting multi-arm spirals (Gerisch (1971)).

There have been speculations --- largely based on Dictyostelium -- that embryonic cells may constitute excitable media through which "positional information" signals propagate, enabling each cell to recognize its position in the whole and to differentiate appropriately (Goodwin and Cohen (1969)). Though no special role for spiral waves has been suggested, it is amusing to note that the navigational scheme most used by airmen today (VHF Omnirange) is based on radio signals with a spiral pattern of audio modulation (Pilot's Handbook (1971)).

It has been argued since 1946 (Wiener and Rosenblueth (1946); Balakhovskii (1965); Krinskii (1968); Gulko and Petrov (1972)) that fibrillation (or perhaps only flutter) in vertebrate hearts is often due to formation of a spiral source near a transient inhomogeneity, as happens in Z reagent and in *Dictyostelium*. A spiral wave would simulate uncoordinated activity, inasmuch as at every moment some locus on the ventricle would be in contraction. Krinskii (1968) showed that spiral cores can be quite small and can *multiply* in heart-like media. The fact that small hearts defibrillate spontaneously (Ruch and Patton (1965)) reminds us that a circular path of circumference approximating the shortest sustainable wavelength is necessary for stable spirals in Z reagent and in *Dictyostelium*. The fact that electric defibrillators work (by synchronizing the whole heart) argues that at least some forms of fibrillation are caused by an unusual but self-sustaining mode of wave propagation, rather than by lesion. But no direct observations of wave geometry have been reported, so far as I know.

It is suggestive that a mathematical model fashioned with the intent to mimic nerve conduction and the heartbeat (Zeeman (1972)) may alternatively describe Z reagent. This model suggests a spiral wave (Winfree (1974)). In fact Gulko and Petrov (1972) have exhibited such a wave by hybrid analog/digital computation using another analog of the Hodgkin-Huxley equations.

Turning to larger-scale events, it seems to me suggestive that the molecular description of Z wave propagation (Field (1974)) recalls ecological interactions by which epidemics and other pest outbreaks are propagated. In fact both periodic recurrences and geographic waves of disease are commonplace phenomena. When we better understand the qualitative features of local kinetics required to support trigger waves, given the right initial conditions, it will be interesting to examine the kinetic models of biogeographic ecology for such features.

WAVE BROOMS

Either kind of wave, like a bucket brigade, can sweep material forward or backward along the direction of propagation. For example, the pseudo-wave of wild-flower-opening that passes up a mountain pasture in the springtime may sweep mutant alleles upward or downward, depending on whether ovaries are more receptive to pollen just before or just after the peak of pollen discharge. Or in a row of cells, suppose membrane permeability to a given substance and intracellular concentration of that substance both vary periodically during the passage of a train of waves at unit velocity. Let permeability vary as sin (x-t) and concentration as sin $(x-t-t_o)$, so the concentration difference across a cell-cell membrane varies as cos $(x-t-t_o)$. The rate of transport between adjacent cells being proportional to the product of permeability by concentration difference, we find a position-independent non-zero average flux per cycle: $J \propto \oint \sin(x-t)\cos(x-t-t_o)dt \propto \sin t_o$. The phase difference t_o determines the direction and rate of transport. A similar mechanism, known to chemical engineers as "parametric pumping" in a reactor bed, has been suggested as a model of active transport (Wilhelm (1966), (1968)). Goodwin (1972 pers. comm.) has discussed another "wavebroom" model in context of coordinate systems for determination of positional information in embryos (Goodwin and Cohen (1969)); substances accumulated by waves near a source of waves could maintain that spot as a source by slightly increasing the local frequency of spontaneous oscillation. Zhabotinsky and Zaikin (1973) may have something of this sort in mind.

WHERE ARE WE?

I have attempted to draw a distinction between "pseudo-waves" having fixed period with variable-to-infinite velocity, and "trigger waves" having fixed velocity and variable-to-infinite period. This is quite likely a parody of reality, there being a continum of intermediate possibilities. These intermediates may occur in Z reagent (and in comparable biological media) near the lower limit of pseudo-wave velocity, where a decelerating pseudo-wave may elicit a trigger wave. This sometimes seems to happen around the meniscus of a dish, from which oxidative activity (blue) spreads inward, at first diffusely and rapidly, then more slowly, until a sharp trigger wave circle begins to propagate inward.

Ortoleva and Ross (1973) may have observed something of this sort in computer simulations of a spatially distributed limit-cycle oscillation which propagated phase waves when the cycle was smooth, but switched to a behavior suggestive of trigger waves when the cycle more resembled a relaxation oscillation.

In pointing to the eigenpatterns of the linearized equation of reaction and diffusion, DeSimone, Beil and Scriven (1973) have drawn attention to waves that are neither pseudo-waves (because the local kinetics has no limit cycle, and in fact can have real eigenvalues) nor trigger waves (because the period between wave arrivals is everywhere the same and is independent of diffusion coefficients). However this model seems unsuitable for Z reagent, since very special initial conditions are implicitly required to avoid exciting the other modes of spatially distributed reaction: this solution has arbitrary amplitude and is only one of a continuum of equally valid eigenpatterns which have different frequencies, wavelengths and velocities and non-zero rate of intensity change (Gmitro and Scriven (1966)). Arbitrary linear combinations are equally valid, e.g. interpenetrating criss-crossed spirals, multi-armed spirals, and standing waves .... none of which occur.

The electrical "slow waves" of intestine (Code and Szurszewski (1970)) also defy classification as "pseudo"- or "trigger"- waves. Diamant, Rose, and Davison (1970) have simulated slow wave behavior using a chain of van der Pol oscillators, each with natural frequency proportional to distance along the intestine, each coupled electrically to its immediate neighbors. The gradient of frequency breaks up into several plateaus of synchrony, within which all oscillators are entrained by the fastest one. This establishes a gradient of phase paralleling the gradient of natural frequency. The wave running down this gradient has the velocity of a pseudo-wave, but its passage is blocked by any transverse cut interrupting the synchronizing interactions. Similar phenomena might appear in a column of Z reagent prepared with a sufficiently steep monotone gradient of natural frequency (Kopell and Howard (1973); Thoenes (1973)).

Limit cycle oscillator models, linear models, and pulse-propagation models (among others, presumably) serve as complementary and overlapping metaphors for the behavior of many biological materials and the new inorganic reagents. However the truest portrait of any particular medium will be painted in colors derived from its particular mechanism. The final problem in all such efforts is, of course, to pursue the equations of mechanism through suitable approximations to an integrated solution, and this is the unchallengeable domain of applied mathematicians. I think the main point of this paper is to suggest a search for propagating, diffusion-dependent solutions with rotational rather than translational invariants: for plane waves first, of course, but then to enquire about the stability of spirals, scrolls, and scroll rings.

BIBLIOGRAPHY

Balakhovskii, I.S. (1965) "Several modes of excitation movement in ideal excitable tissue" Biofizika 10, 1063-1067 (1175-1179 in English translation).

Beck, M.T., Varadi, Z.B. (1972) "One, two and three-dimensional spatially periodic chemical reactions" Nature Physical Science 235, 15-16.

Belousov, B. (1958) Sb. Ref. Radiats. Med. (collection of abstracts on radiation medicine) (Medgiz, Moscow).

Busse, H.G. (1969) "A spatial periodic homogeneous chemical reaction" J. Phys. Chem. 73, 750.

Code, F. and Szurszewski, H. (1970) "The effect of duodenal and mid small bowel transection on the frequency gradient of the pacesetter potential in the canine small intestine" J. Physiol. 207, 281-289.

Cohen, H. (1972) "Models of clocks and maps in developing organisms" from Some Mathematical Questions in Biology". II ed. Jack Cowan; 3-32 (Am. Math. Soc. Providence, R.I.)

Cohen, M.H. and Robertson, A. (1971) "Wave propagation in the early stages of aggregation of cellular slime molds" J. Theor. Biol. 31, 101-118.

DeSimone, J.A., Beil, D.L., Scriven, L.E. (1973) "Ferroin-collodion membranes: dynamic concentration patterns in planar membranes" Sci. 180, 946-948.

Diamant, N.E., Rose, T.K., Davison, E.J. (1970) "Computer simulation of intestinal slow-wave frequency gradient" Am. J. Physiol. 29, 1684-1690.

Durston, A.J. (1974) "Dictyostelium discoideum aggregation fields as excitable media" J. Theor. Biol. (in press).

Durston, A.J. "Pacemaker activity during aggregation in Dictyostelium discoideum" (in preparation)

Field, R.J. (1972) "A reaction periodic in space and time" J. Chem. Ed. 49, 308-311.

Field, R.J., Koros, E. and Noyes, R.M. (1972) "Oscillations in chemical systems. II. Thorough analysis of temporal oscillation in the bromate-cerium-malonic acid system" J. Amer. Chem. Soc. 94, 8649-8664.

Field, R.J. and Noyes, R.M. (1972) "Explanation of spatial band propagation in the Belousov reaction" Nature 237, 390-392.

Field, R.J. (1974) "The generation of spatial and temporal structure by a chemical reaction" Chemie in Unserer Zeit (in press).

Field, R.J. and Noyes, R.M. (1974b) "Oscillations in chemical systems. IV. limit cycle behavior in a model of a real chemical reaction" (in press).

Gerisch, G. (1965) "Stadienspezifische Aggregationsmuster bei Dictyostelium discoideum" Wilhelm Roux Archiv Entwicklungsmech. Organismen 156, 127.

Gerisch, G. (1968) "Cell aggregation and differentiation in Dictyostelium" In Current Topics in Developmental Biology 3, 157-197. A. Moscona and A. Monroy, eds. (Academic Press, New York).

Gerisch, G. (1971) "Periodische Signale steuren die Musterbildung in Zellverbanden" Naturwissenschaften 58, 430-438.

Glansdorff, P. and Prigogine, I. Thermodynamic Theory of Structure, Stability, and Fluctuations (Wiley, London, 1971).

Gmitro, J.I. and Scriven, L.E. (1966) "A physicochemical basis for pattern and rhythm" pp. 221-255 of Intracellular Transport J. Danielli ed. (Academic Press, New York).

Goodwin, B.C. and Cohen, M.H. (1969) "A phase-shift model for the spatial and temporal organization of developing systems" J. Theoret. Biol. 25, 49-107.

Gul'ko, F.B. and Petrov, A.A. (1972) "Mechanism of formation of closed pathways of conduction in excitable media" Biophysics 17, 271-281.

Herschkowitz-Kaufman, M. (1970) "Structures dissipatives dans une reaction chemique homogene" C.R. Acad. Sc. Paris 270, 1049.

Herschkowitz-Kaufman, M. and Nicolis, G. (1972) "Localized spatial structures and nonlinear chemical waves in dissipative systems" J. Chem. Phys. 56, 1890-1895.

Hodgkin, A. L. and Huxley, A. F. (1952) "A quantitaive description of membrane current and its application to conduction and excitation in nerve" J. Physiol. 117, 500-544.

Kopell, N. and Howard, L. N. (1973) "Horizontal bands in the Belousov reaction" Sci. 180, 1171-1173.

Krinskii, V.I. (1968) "Fibrillation in excitable media" Probl. Kibernetiki 20, 59-80.

Ortoleva, P. and Ross, J. (1973) "Phase waves in oscillatory chemical reactions" J. Chem. Phys. 58, 5673-5680.

Pilots Handbook of Aeronautical Knowledge (1971): Federal Aviation Administration Washington, D.C.

Rinzel, J. and Keller, J.B. (1973) "Travelling wave solutions of a nerve conduction equation" Biophys. J. 13, 1313-1337.

Robertson, A. (1972) "Quantitative analysis of the development of cellular slime molds" from Some Mathematical Questions in Biology III. Jack Cowan ed. 48-73. (Am. Math. Soc., Providence, R.I.)

Robertson, A., Drage, D.J., Cohen, M.H. (1972) "Control of aggregation in Dictyostelium discoideum by an external periodic pulse of cyclic adenosine monophosphate" Sci. 175, 333-335.

Ruch, T.C. and Patton, H.D. (1965) Physiology and Biophysics (Saunders, Phila.) p. 587.

Smoes, M. and Dreitlein, J. (1973) "Dissipative structures in chemical oscillations with concentration-dependent frequency" J. Chem. Phys. (in press).

Tatterson, D.F. and Hudson, J.L. (1973) "An experimental study of chemical wave propagation" Chem. Eng. Commun. 1, 3-11.

Thoenes, D. (1973) " 'Spatial oscillations' in the Zhabotinsky reaction" Nature Phys. Sci. 243, 18-21.

Wiener, N. and Rosenblueth, A. (1946) "The mathematical formulation of the problem of conduction of impulses on a network of connected excitable elements, specifically in cardiac muscle" Arch. Inst. Cardiologia de Mexico 16, (3) 105.

Wilhelm, R.H. (1968) "Parametric pumping" Ind. Eng. Chem. Funda. 7, 337.

Wilhelm, R.H. (1966) "Parametric pumping - a model for active transport" in Intracellular Transport J. Danielli ed. (Academic Press, N.Y.)

Winfree, A.T. (1970) "Oscillatory control of cell differentiation in Nectria?" Proc. I.E.E.E. Symp. Adapt. Process 9, pp. 23.4.1-23.4.7.

Winfree, A.T. (1970b) "The temporal morphology of a biological clock" in Lectures on Mathematics in the Life Sciences vol. 2, M. Gerstenhaber ed. pp. 109-150. (Am. Math. Soc., Providence, R.I.).

Winfree, A.T. (1972) "Spiral waves of chemical activity" Sci. 175, 634-636.

Winfree, A.T. (1973) "Polymorphic pattern formation in the fungus Nectria" J. Theor. Biol. 38, 363-382.

Winfree, A.T. (1973b) "Scroll-shaped waves of chemical activity in three dimensions" Sci. 181, 937-939.

Winfree, A.T. (1974) "Spatial and temporal organization in the Zhabotinsky reaction" in Aharon Katchalsky Memorial Symposium Science and Humanism: Partners in Human Progress H. Mel ed. (University of California Press).

Zhabotinsky, A.M. and Zaikin, A.N. (1970) "Concentration wave propagation in two dimensional liquid-phase self-oscillating systems" Nature 225, 535-537.

Zhabotinsky, A.M. and Zaikin, A.N. (1973) "Auto-wave processes in a distributed chemical system" J. Theor. Biol. 40, 45-61.

Zeeman, E.C. (1972) "Differential equations for the heartbeat and nerve impulse" Towards a Theoretical Biology Vol. 4, C.H. Waddington ed. pp. 8-67.

MATHEMATICS AND BIOLOGY*

P. Rajagopal

Professor of Computer Science & Mathematics

Atkinson College, York University, Downsview

Ontario, Canada

Science tries to deal with reality and even the more precise sciences may be based on poorly understood approximations toward which the scientist must maintain an appropriate scepticism. Thus, for instance, it may come as a shock to the mathematician to learn that the Schrodinger equation for the hydrogen atom, which he is able to solve only after a considerable effort of functional analysis and special function theory, is not a literally correct description of this atom, but only an approximation to a somewhat more correct equation taking account of spin, magnetic dipole, and relativistic effects; that this corrected equation is itself only an ill-understood approximation to an infinite set of quantum field-theoretical equations; and finally that the quantum field theory, besides diverging, neglects a myriad of strange-particle interactions whose strength and form are largely unknown. The physicist, looking at the original Schrodinger equation, learns to sense in it the presence of many invisible terms, integral, integrodifferential, perhaps even more complicated types of operators, in addition to the differential terms visible, and this sense inspires an entirely appropriate disregard for the purely technical features of the equation which he sees.

---

*Based, in part, on a discussion chaired by Dr. A. B. Tayler.

This very healthy self-scepticism is foreign to the axiomatic mathematical approach.

Mathematics must deal with well-defined situations. Thus, in its relations with science, mathematics depends on an intellectual effort outside of mathematics for the crucial specification of the approximation which mathematics is to take literally. Give a mathematician a situation which is at all ill-defined, then he must first make it well defined. Perhaps appropriately, but perhaps also inappropriately. The hydrogen atom is a good example of this process. The physicist asks: 'What are the eigenfunctions of such-and-such a differential operator?' The mathematician replies: 'The question as posed is not well defined. First you must specify the linear space in which you wish to operate, then the precise domain of the operator as a subspace. Carrying all this out in the simplest way, we find the following result . . .' Whereupon the physicist may answer, much to the mathematician's chagrin: 'Incidentally, I am not so much interested in the operator you have just analyzed as in the following operator, which has four or five additional small terms -- how different is the analysis of this modified problem?' In this example one may perhaps consider that nothing much is lost, nothing at any rate but the vigor and wide sweep of the physicist's less formal attack. But, in other cases, the mathematician's habit of precision may have more unfortunate consequences. The mathematician turns the scientist's theoretical assumptions, i.e., convenient points of analytical emphasis, into axioms, and then takes these axioms literally. This brings with it the danger that he may also persuade the scientist to take these axioms literally. The question, central to the scientific investigation but intensely disturbing in the mathematical context -- what happens to all this if the axioms are relaxed? -- is thereby put into shadow.

The exact nature of mathematics thus makes it essential, if mathematics is to be appropriately used in science, that the assumptions upon which mathematics is to elaborate be correctly chosen from a larger point of view, invisible to mathematics itself. The single-mindedness of mathematics reinforces this conclusion. Mathematics is able to deal successfully only with the simplest of situations, more precisely, with a complex situation only to the extent that rare good fortune makes this complex situation hinge upon a few dominant simple factors. Beyond the well-traversed path, mathematics loses its bearings in a jungle of unnamed special functions and impenetrable combinatorial particularities. Thus, the mathematical technique can only reach far if it starts from a point close to the simple essentials of a problem which has simple essentials. That form of wisdom which is the opposite of single-mindedness, the ability to keep many threads in hand, to draw for an argument from many disparate sources, is quite foreign to mathematics. Only with difficulty does it find its way to the scientist's ready grasp of the relative importance of many factors. Quite typically, science leaps ahead and mathematics plods behind. Hence, the formulation of a problem in mathematical terms is difficult, and requires that mathematicians and biologists attempt to communicate with each other.

Related to this and perhaps more productive of rueful consequence, is the simple-mindedness of mathematics, its willingness, like that of a computing machine, to elaborate upon any idea, however absurd; to dress scientific brilliancies and scientific absurdities alike in the impressive uniform of formulae and theorems. Unfortunately however, an absurdity in uniform is far more persuasive than an absurdity unclad. The very fact that a theory appears in mathematical form, that for instance, a theory has provided the occasion for the application of a fixed-point theorem, or of a result about difference equations, somehow makes us more ready to take it seriously. And the mathematical-intellectual effort of applying the theorem fixes in us the particular point of view of the theory with which we deal, making us blind to whatever appears neither as a dependent nor as an independent parameter in its mathematical formulation. The result, perhaps most common in the social sciences, is bad theory with a mathematical passport.

We now turn to mathematical education. It is an unfortunately common experience that an undergraduate mathematics major, having had mathematics in school through college is left with a false impression of mathematics. He may have little sense of mathematics as a part of culture, nor know that it is not complete or final, that new areas beckon him for work and old ones for improvement. What is required of a student is not so much command of the various techniques but the quality celebrated in the ideology of mathematical education, namely "mathematical maturity". This quality requires mathematical experience in as wide an area as possible, and is not merely a collection of techniques.

Any attempt to apply mathematics to the world involves three stages. First we observe the phenomena and formulate a mathematical description in the form of a differential equation, algebraic equation or whatever. We then temporarily forget the real world and use mathematical reasoning to solve the equation. This stage may involve inventing new mathematics or extending what exists. Finally we return to the real world and interpret this solution in terms of reality, this interpretation may require experimental testing. Commonly the most difficult stage is the first one, this is certainly so in biology at present since we rarely know enough about the "laws" governing the components of biological systems to write down their appropriate relationships with confidence. One tries to draw analogies with physical sciences but one should not overlook the fact that the laws of physical sciences have been clear since the work of Newton. The problems in physical sciences today are not the construction of a proper model; the known physical laws have dictated the model.

The difficult problems of physical science are typically those of developing ingenious methods of solving complex equations. For instance, the rapid development of analysis in the eighteenth and nineteenth centuries was spurred by problems arising in celestial and terrestrial mechanics, in the propagation of waves associated with theories of sound, heat, light and electro-magnetic phenomena. The whole menagerie of special functions arose in the process of serving the needs of those areas of investigations, such as perturbation theory in celestial mechanics, elasticity theory, statistical mechanics, etc. Not the least important of these developments was that of numerical analysis, a highly valued skill in the days when electronic computers were unknown. One can recall the prodigious feat of Gauss in relocating, at the cost of monumental calculations, the asteriod Ceres after it had been "lost". The situation in the non-physical sciences is now reversed. The electronic computer can do in hours what it took Gauss decades to do, so that the complexity of mathematical models is not the principal constraint. The main issue is the propriety of the mathematical model.

It is against this critical background of the nature of mathematical activity, mathematical education and use of mathematics in applications, one can appreciate the Conference. As one who has been working in mathematics and numerical analysis for several years the Conference provided me an excellent opportunity to talk to users of mathematics in biological problems. Conferences of this type, where a large range of biological work is represented would provide the best introduction to undergraduates in mathematics or others associated with mathematical activity who wish to learn about mathematics in biology or desire to become involved in biological work. Most of the participants at the discussion meeting were in rough agreement with the ideas on mathematical relevance raised above.

The teaching of mathematics to ecologists was also discussed. Maynard Smith's book was mentioned quite frequently as being one of the best introductions. Statistical methods were mentioned to be of great use too, but theoretical courses spending much of their time on hypothesis testing based on assumptions of normality seemed to be of much less importance than the nature of observations and measurement and the quality of resultant data. Much of the data available did not seem to conform precisely to the requirements of any one model. In fact, the data often seems to be appropriate for use with two or three models, so that deciding on the model that is most appropriate is often simply a matter of individual judgment, requiring biological insight and leading to disagreement among the biologists.

Theory construction and model building are now essential for the practising biologist. Courses on this topic may well be of great value. In a course of this type, a student should have as much experience as possible in constructing models. The problems chosen have to be realistic; quite often it may be necessary to construct a succession of models in an effort to find a satisfactory one. The course should make the student aware that there may be several approaches which lead to essentially different mathematical models for the same problem. Therefore a critical evaluation of the steps in constructing a model is essential in order that the student know what kind of information he can expect or cannot expect from a model and that he be able to choose the model which is most effective to the purpose. Epidemic models (deterministic or stochastic), oscillation models, differential equation models for wave propagation, and kinetic models described in this volume could be used for bringing out the powers and limitations of modelling in such a course.

Mathematics is a unifying language; mathematical descriptions are of descriptive and pedagogic value. Hence they can help insight and understanding. There is therefore need to teach limitations of tools as much as tools. It is an open question whether mathematicians can do this on their own. Lieberman's collection of articles is an example of such an attempt in statistics. The possibility of a book for biology similar to that of Ben Noble was mentioned, and it was observed that the Mathematical Association of America had at one time started such a venture.

The meaning of mathematics is more in its process than in its product, but typically mathematics knows better what to do than why to do it. Involved mathematical argument tends to hide the fact that we understand only poorly what it is based on. Biologists should therefore ask what kind of numerate biologists can do biology well and acquire the skills themselves before teaching them. The exceptional individuals teach themselves, but it is institutionalised methods which pave the way for larger numbers to carry on the work.

Cowan, during his comprehensive review of large scale nervous activity mentioned a sign on his table which reads: "Dear Lord, please make the world linear, stationary, Gaussian and non-threshold". Perhaps, therein lies a summary of what awaits a numerate biologist. The world seems to prefer non-linearity, non-stationarity, non-Gaussianity and threshold methods. The challenge of education lies in providing opportunities for the emergence of Eulers, Lagranges, Schrodingers and Wieners who could provide and evolve appropriate methods for biology (and mathematics) as their predecessors did for physical sciences (and mathematics).

REFERENCES:

Lieberman, B. (Ed.)
Contemporary problems in statistics
Oxford University Press, Toronto (1971)

Noble, B.
Applications of undergraduate mathematics in engineering
Mathematical Association of America (1967)
Macmillan, New York

Smith, J. M.
Mathematical ideas in biology
Cambridge University Press (1968)

SOME PROBLEMS POSED AT THE CONFERENCE

Problem 1

Suppose there are N species in an environment.

Suppose that there has been taken m samples, $S_1, S_2, \ldots, S_m$. Each sample contains from 0 to N species. Let $D_i(x)$ be the probability distribution for the number x of samples taken before encountering the first representative of species i.

Make some assumptions about the D_i (e.g. independent negative binomial distributions) and devise an estimating procedure to estimate the true N from knowledge of the species represented in each of the m samples.

More generally, given the species represented in the samples, devise a model under whatever assumptions seem reasonable, interesting, or appropriate, from which an estimate of the number of species, N, in the environment can be made.

Problem 2

For a finite set S , $d : S \times S \to R+$ is a real non-negative function with $d(a, b) = 0$ iff $a = b$. d may be considered a measure of difference, for pairs of species in S .

Suppose $T \supseteq S$, $T \times T \supseteq E$, the edges of a non-directed tree graph with vertices T ; i.e. $E^{-1} = E$.

Let $q : E \to R+$ be a non-negative function of the edges in E .

For any pair (s, s') of elements in S , let P be the collection of edges in E constituting the unique path from s to s' . Define

$$p(s, s') = \sum_{e \in P} q(e) .$$

We seek, for any given S and d , that triple (T, E, q) described above such that

$$\sum_{(s, s') \in S \times S} |p(s, s') - d(s, s')|$$

is a minimum.

Solution conforms to that estimate of evolutionary history "most compatible" with the idea that phenetic and patristic evolutionary distances are most nearly equal.

Problem 3

Let S be a finite collection of points in the positive orthant of an n dimensional Euclidean Space. Find a finite subset T containing S and the origin, together with a directed tree partial order on T subject to the following constraints:

1. The origin is the root of T.
2. Let C be the corresponding cover relation; for tCt' read t covers t'.

   Write $t = (t_1, t_2, \ldots, t_n)$, $t' = (t'_1, t'_2, \ldots, t'_n)$. Constraint 2 requires that if tCt' then $t_i \geq t'_i$ for $i = 1,2,\ldots,n$.
3. For tCt' define $d(t, t') = \sum_{i=1}^{n} (t_i - t'_i)$.

   Define $L(C, T) = \sum_{(t, t') \in C} d(t, t')$.

   Constraint 3 requires that if (C', T') is any directed tree partial order for any set $T' \supseteq S$, then $L(C, T) \leq L(C', T)$.

A practical solution is an efficient computer algorithm. Solution conforms with that sought by Camin-Sokal Model of J. Theoret. Biol. 21 : 421-438 (1968).

G. F. Estabrook
Department of Botany
University of Michigan
Ann Arbor, Michigan 48104

THREE-DIMENSIONAL PHASE SPACE

Richard C. Hertzberg

Center for Quantitative Science
in Forestry, Fisheries and Wildlife
University of Washington
Seattle, Washington 98195

Problem 4

In the study of sustained oscillations in enzyme systems, the simplest case involves three chemical components with feedback. The rate equations give rise to a system of three ordinary nonlinear differential equations, denoted in matrix form,

$$\dot{\underline{X}} = \underline{f}(\underline{X}), \ \underline{X} = \begin{bmatrix} x_1 \\ x_2 \\ x_3 \end{bmatrix}.$$

Since solving such a system for a closed (e.g., limit cycle) solution is usually difficult, it is desirable to reduce it to a 2-dimensional problem, where theorems abound. From the 3-dimensional system one can develop three 2-dimensional phase-plane projections of the solution. If each 2-dimensional projection is a convex closed curve (periodic), with nonzero curvature of constant sign, is this sufficient to make the 3-dimensional phase-space trajectory a planar closed curve? If not, what other conditions are needed? E.g., continuous derivatives; no points of zero motion, i.e., nowhere does $dx_i/dt = 0$ is the $x_i x_j$ phase plane. Counterexamples if zero curvature is allowed are a four-sided polygon (A) and the curve formed by connecting the midpoints with quarter-circles (B).

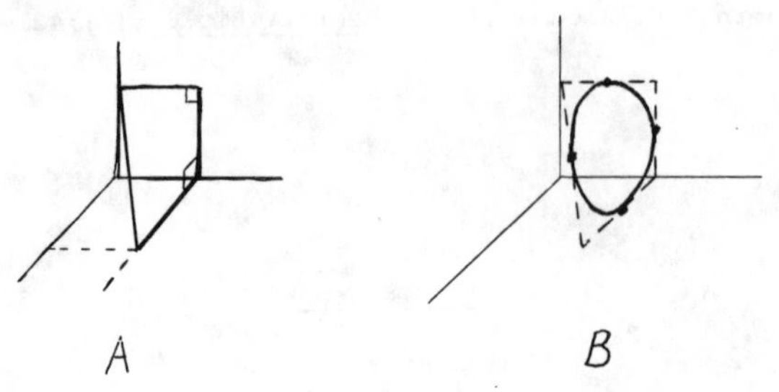

A BIBLIOGRAPHY OF MATHEMATICAL BIOLOGY

George W. Swan
Department of Pure and Applied Mathematics
Washington State University
Pullman, Washington 99163, U.S.A.

This bibliography was compiled by the author during the spring of 1973 and was distributed to students attending a graduate level course in biomathematics. References to areas in biophysics (for example ion transport across real and synthetic membranes), cell biology (growth, function and form) and the nervous system are not included here; a more comprehensive bibliography would, of course, need to include these areas. The bibliography does not contain references to scientific papers. The listings are by year of publication, the oldest publications being at the top and the latest are at the bottom. Within each year the listings are by author alphabetically.

TOPICS IN GENERAL MATHEMATICAL BIOLOGY

Rashevsky, N., Advances and Applications of Mathematical Biology, University of Chicago Press, Chicago, 1940.

Rashevsky, N., Mathematical Biophysics, (revised edition), The University of Chicago Press, Chicago, 1948.

Bailey, N. T. J., The Mathematical Theory of Epidemics, Griffin, London, 1957.

Rashevsky, N., Mathematical Biophysics, Physico-Mathematical Foundations of Biology; Dover Publications, New York, Vol. 1, Vol. 2, 1960. (This is a revision of the above work by Rashevsky.)

Rashevsky, N., Mathematical Principles in Biology and Their Applications, Charles C. Thomas, Springfield, 1961.

Lucas, H. L., Cullowhee Conference on Training in Biomathematics, Typing Service, North Carolina State University, Raleigh, 1962. (Copies available from Institute of Statistics at $5.25.)

Mathematical Problems in the Biological Sciences, Proceedings of Symposia in Applied Mathematics Vol. XIV, American Math. Soc., Providence, 1962.

Attinger, E. O., (editor), International Symposium on Pulsatile Blood Flow, 1st, Philadelphia, 1963, Blakiston Division of McGraw-Hill, New York, 1964.

Bailey, N. T. J., The Elements of Stochastic Processes with Applications to the Natural Sciences, John Wiley and Sons, New York, 1964.

Tentative recommendations for the undergraduate program of students in the biological management and social sciences, 1964. (CUPM Central Office, P.O. Box 1024, Berkeley, CA, 94701.)

Waterman, T. H., and Morowitz, H. J. (editors), Theoretical and Mathematical Biology, Blaisdell, New York, 1965.

Batschelet, E., Statistical Methods for the Analysis of Problems in Animal Orientation and Certain Biological Rhythms, American Institute of Biological Sciences, Washington, D. C., 1965.

Sollberger, A., Biological Rhythm Research, Elsevier Publishing Co., New York, 1965.

Bailey, N. T. J., The Mathematical Approach to Biology and Medicine, J. Wiley and Sons, New York, 1967.

Rosen, R., Optimality Principles in Biology, Butterworths, London, 1967.

Baker, J. T. W., and Allen, G. E., Hypothesis, Prediction and Implication in Biology, Addison-Wesley, Reading, 1968.

Biokybernetik, Band 1, Materialien des 1. Internationalen Symposiums, "Biokybernetik", Leipzig, Sept., 1967, Karl-Marx-Universität, Leipzig, 1968.

Biokybernetik, Band 2, Materialen des 1. Internationalen Symposiums, "Biokybernetik", Leipzig, Sept., 1967, Karl-Marx-Universität, Leipzig, 1968.

Smith, J. M., Mathematical Ideas in Biology, Cambridge University Press, Cambridge, 1968.

Symposium on Mathematical Biology, 1st., Washington D. C., 1966: Some Mathematical Problems in Biology, Vol. 1, American Math. Soc., Providence, 1968.

Heinmets, F. (editor), Concepts and Models of Biomathematics, Vol. 1, M. Dekker, New York, 1969.

Hoffman, W. C., Bibliography for an Undergraduate Course in Applications of Mathematics in the Life Sciences, Dec., 1970. (CUPM Central Office, P.O. Box 1024, Berkeley, CA, 94701.)

The Life Sciences, Recent Progress and Application to Human Affairs, The World of Biological Research, Requirements for the Future, National Academy of Sciences, Washington, D. C., 1970.

Luce, G. G., Biological Rhythms in Human and Animal Physiology, U. S. Department of Health, Education and Welfare, and National Institute of Mental Health, 1970.

Recommendations for the Undergraduate Mathematics Program for Students in the Life Sciences, Sept., 1970. (CUPM Central Office, P.O. Box 1024, Berkeley, CA, 94701.)

Smith, R. E., Biorhythm Theory, Control Data Institute, Control Data Corporation, Minneapolis, 1970.

Symposium on Mathematical Biology, 2nd., New York, 1967; and Symposium on Mathematical Biology, 3rd., Dallas, 1968: Some Mathematical Questions in Biology, Vol. 2, American Math. Soc., Providence, 1970.

Colquhoun, W. P. (editor), Biological Rhythms and Human Performance, Academic Press, New York, 1971.

Gel'fand, I. M., Gurfinkel, V. S., Fomin, S. V., and Tsetlin, M. L. (editors), Models of the Structural-Functional Organization of Certain Biological Systems, M.I.T. Press, Cambridge, 1971.

Mathematical Aspects of Life Sciences (Proceedings of the Symposium held at Queen's University in June, 1969) edited by M. T. Wasan, Queen's papers in pure and applied mathematics no. 26, Queen's University, Kingston, Ontario, 1971.

Rosen, R. (editor), Foundations of Mathematical Biology, Vol. 1: Subcellular Systems, Academic Press, New York, 1972.

Rosen, R. (editor), Foundations of Mathematical Biology, Vol. 2: Cellular Systems, Academic Press, 1972.

Rosen, R. (editor), Foundations of Mathematical Biology, Vol. 3: Supercellular Systems, Academic Press, 1972.

Welch, J. C., Potchen, E. J., and Welch, M. J., Fundamentals of the Tracer Method, W. B. Saunders, Philadelphia, 1972.

Symposium on Mathematical Biology, 4th., Boston, 1969: Some Mathematical Questions in Biology II, Vol. 3, American Math. Soc., Providence, 1972.

Symposium on Mathematical Biology, 5th., Chicago, 1970: Some Mathematical Questions in Biology III, Vol. 4, American Math. Soc., Providence, 1972.

Lieberstein, H. M., Mathematical Physiology, Blood Flow and Electrically Active Cells, American Elsevier, 1973.

Rubinow, S., Mathematical Problems in the Biological Sciences, (CBMS Regional Conference Series in Applied Mathematics), SIAM Publications, 33 S. 17th St., Philadelphia, 1973.

MATHEMATICAL TECHNIQUES IN BIOLOGY, MEDICINE

Feldman, W. M., Biomathematics, Charles Griffin and Co., London, 1923.

Kostitzin, V. A., Mathematical Biology, Harrap, London, 1939.

Saunders, L., and Fleming, R., Mathematics and Statistics for Use in Pharmacy, Biology and Chemistry, Pharmaceutical Press, London, 1957.

Defares, J., and Sneddon, I. N., The Mathematics of Medicine and Biology, North Holland, Amsterdam, 1961. (Also, Year Book Medical Publishers, Chicago, 1961.)

Moran, P. A. P., The Statistical Processes of Evolutionary Theory, Clarendon Press, Oxford, 1962.

Nahikian, H. M., A Modern Algebra for Biologists, University of Chicago Press, Chicago, 1964.

Searle, S. R., Matrix Algebra for the Biological Sciences, Including Applications in Statistics, J. Wiley and Sons, New York, 1966.

Smith, C. A. B., Biomathematics, Vol. 1, (4th edition) Hafner Publ. Co., New York, 1966.

Stibitz, G. B., Mathematics in Medicine and the Life Sciences, Year Book Medical Publishers, Inc., Chicago, 1966.

Rosen, R., Optimality Principles in Biology, Butterworths, London, 1967.

Bartlett, M. S., Biomathematics, Clarendon Press, Oxford, 1968.

Chiang, C. L., Introduction to Stochastic Processes in Biostatistics, J. Wiley and Sons, New York, 1968.

Mosimann, J. E., Elementary Probability for the Biological Sciences, Appleton-Century-Crofts, New York, 1968.

Crowe, A., and Crowe, A., Mathematics for Biologists, Academic Press, New York, 1969.

Rosen, R., Dynamical System Theory in Biology, Vol. 1: Stability Theory and its Applications, J. Wiley and Sons, New York, 1970.

Batschelet, E., Introduction to Mathematics for Life Scientists, Springer-Verlag, New York, 1971.

Colquhoun, D., Lectures on Biostatistics; An Introduction to Statistics with Applications in Biology and Medicine, Clarendon Press, Oxford, 1971.

Schadach, D. J., Biomathematik, I, Kombinatorik, Wahrscheinlichkeit und Information, Akademie-Verlag, Berlin, 1971.

Schadach, D. J., Biomathematik, II, Graphen, Halbgruppen und Automaten, Akademie-Verlag, Berlin, 1971.

Ashton, W. D., The Logit Transformation (with special reference to its uses in Bioassay), Griffin, London, 1972.

Bahn, A., Basic Medical Statistics, Grune and Stratton, New York, 1972.

Simon, W., Mathematical Techniques for Physiology and Medicine, Academic Press, New York, 1972.

Sokal, R. R., and Rohlf, F. J., Introduction to Biostatistics, W. H. Freeman, San Francisco, 1972.

Vann, E., Fundamentals of Biostatistics, D. C. Heath, Lexington, 1972.

Baxter, W. E., and Sloyer, C. W., Calculus with Probability for the Life and Management Sciences, Addison-Wesley, Reading, 1973.

POPULATION STUDIES

Volterra, V., Lecons sur la Théorie Mathématique de la Lutte pour la Vie, Gauthier-Villars, Paris, 1931.

Ullyett, G. C., Biomathematics and Insect Population Problems--a Critical Review. Memoir Entomol. Soc. of S. Africa, Pretoria. No. 2, 1953.

Lotka, A. J., Elements of Mathematical Biology, Dover, New York, 1956.

Bartlett, M. S., Stochastic Population Models in Ecology and Epidemiology, Methuen and Co., London, 1960.

Bailey, N. T. J., Introduction to the Mathematical Theory of Genetic Linkage, Clarendon Press, Oxford, 1961.

Keyfitz, N., Introduction to the Mathematics of Population, Addison-Wesley, Reading, 1968.

Kuczynski, R. R., The Measurement of Population Growth: Methods and Results, Demographic Monographs, Vol. 6, Gordon and Breach, New York, 1969.

Ehrlich, P. R., and Ehrlich, A. H., Population Resources Environment, Issues in Human Ecology, W. H. Freeman, San Francisco, 1970.

Kojima, K., (editor), Biomathematics, Vol. 1, Mathematical Topics in Population Genetics, Springer-Verlag, New York, 1970.

Andrewartha, H. G., Introduction to the Study of Animal Populations, (2nd edition), Methuen, London, 1971.

Cavalli-Sforza, L. L., and Bodmer, W. F., The Genetics of Human Populations, W. H. Freeman, San Francisco, 1971.

Goel, N. S., Maitra, S. C., and Montroll, E. W., On the Volterra and Other Nonlinear Models of Interacting Populations, Academic Press, New York, 1971.

Keyfitz, N., and Flieger, W., Population: Facts and Methods of Demography, W. H. Freeman, San Francisco, 1971.

Wilson, E. O., and Bossert, W. H., A Primer of Population Biology, Sinauer Associates, Stanford, 1971.

The Mathematical Theory of the Dynamics of Biological Populations, (Conference), Oxford; The Institute of Mathematics and its Applications, Southend-on-Sea, Essex, England, 1972.

Cole, H. S. D., et al., (editors), Models of Doom: A Critique of the Limits to Growth, Universe Books, New York, 1972.

Greville, T. N. E., (editor), Population Dynamics, Proceedings of the Symposium on Population Dynamics conducted by the Mathematics Research Center, University of Wisconsin, Madison, June 19-21, 1972, Academic Press, New York, 1972.

Meadows, D. H., Meadows, D. L., Randers, J., and Behrens III, W. W., The Limits to Growth, (A Report for the Club of Rome's Project on the Predicament of Mankind), Universe Books, New York, 1972.

Williamson, M., The Analysis of Biological Populations, E. Arnold, London, 1972.

THERMODYNAMIC CONSIDERATIONS

Katchalsky, A., and Curran, P. F., Nonequilibrium Thermodynamics in Biophysics, Harvard University Press, Cambridge, 1965.

Linford, J. H., An Introduction to Energetics with Applications to Biology, Butterworths, London, 1966.

Prigogine, I., Introduction to Thermodynamics of Irreversible Processes (third edition), Interscience Publishers, New York, 1967.

Morowitz, H. J., Energy Flow in Biology; Biological Organization as a Problem in Thermal Physics, Academic Press, New York, 1968.

Morowitz, H. J., Entropy for Biologists; An Introduction to Thermodynamics, Academic Press, New York, 1970.

Glandsdorff, P., and Prigogine, I., Thermodynamic Theory of Structure, Stability and Fluctuations, Wiley-Interscience, New York, 1971.

CONTROL THEORIES

Mesarović, M. D., The Control of Multivariable Systems, M.I.T. Press, Cambridge, 1960.

Grodins, F. S., Control Theory and Biological Systems, Columbia University Press, New York, 1963.

Riggs, D. S., The Mathematical Approach to Physiological Problems, The Williams and Wilkins Co., Baltimore, 1963.

Milsum, J. H., Biological Control Systems Analysis, McGraw-Hill, New York, 1966.

DiStefano, J. J., Stubberud, A. R., and Williams, I. J., Feedback and Control Systems, Schaum Outline Series, McGraw-Hill, New York, 1967.

Mesarović, M. D. (editor), Systems Theory and Biology, Springer-Verlag, New York, 1968.

Milsum, J. H., Positive Feedback; a General Systems Approach to Positive/Negative Feedback and Mutual Causality, 1968.

Mesarović, M. D., Macko, D., and Takahara, Y., Theory of Hierarchical, Multilevel Systems, Academic Press, New York, 1970.

Riggs, D. S., Control Theory and Physiological Feedback Mechanisms, The Williams and Wilkins Co., Baltimore, 1970.

Riggs, D. S., The Mathematical Approach to Physiological Problems: A Critical Primer, M.I.T. Press, Cambridge, 1970.

Hassenstein, B., Information and Control in the Living Organism, an elementary introduction, Chapman and Hall, London, 1971.

TAXONOMY

Gregg, J. R., The Language of Taxonomy, an application of symbolic logic to the study of classificatory systems, Columbia University Press, New York, 1954.

Sokol, R. R., and Sneath, P. H. A., Principles of Numerical Taxonomy, W. H. Freeman, San Francisco, 1963.

Jardine, N., and Sibson, R., Mathematical Taxonomy, John Wiley and Sons, New York, 1971.

THE BRAIN

von Neumann, J., The Computer and the Brain, Yale University Press, New Haven, 1958.

Arbib, M., Brains, Machines and Mathematics, McGraw-Hill, New York, 1964.

Kabrisky, M., A Proposed Model for Visual Information Processing in the Human Brain, University of Illinois Press, Urbana, 1966.

Stevens, C. F., Neurophysiology: a primer, J. Wiley and Sons, New York, 1966.

Eccles, J. C., The Understanding of the Brain, McGraw-Hill, New York, 1972.

ECOLOGY

Kershaw, K. A., Quantitative and Dynamic Ecology, Edward Arnold, London, 1964.

Pattee, H. H., Edelsack, E. A., Fein, L., and Callahan, A. B., Natural Automata and Useful Simulations, Spartan Books, Washington, D. C., 1966.

Watt, K. E. F., (editor), Systems Analysis in Ecology, Academic Press, New York, 1966.

Watt, K. E. F., Ecology and Resource Management: A Quantitative Approach, McGraw-Hill, Hightstown, New Jersey, 1968.

Pielou, E. C., An Introduction to Mathematical Ecology, Wiley-Interscience, New York, 1969.

Chaston, I., Mathematics for Ecologists, Butterworths, London, 1971.

Patil, G. P, Pielou, E. C., and Waters, W. E., (editors), Statistical Ecology, Vol. 1: Spatial Patterns and Statistical Distributions, Pennsylvania State University Press, University Park, 1971.

Patil, G. P., Pielou, E. C., and Waters, W. E., (editors), Statistical Ecology, Vol. 2: Sampling and Modeling Biological Populations and Population Dynamics, Pennsylvania State University Press, University Park, 1971.

Patil, G. P., Pielou, E. C., and Waters, W. E., (editors), Statistical Ecology, Vol. 3: Many species Populations, Ecosystems and Systems Analysis, Pennsylvania State University Press, University Park, 1971.

Patten, B. C. (editor), Systems Analysis and Simulation in Ecology, Vol. 1, Academic Press, New York, 1971.

Ehrlich, P. R., Ehrlich, A. E., and Holdren, J. P., Human Ecology: Problems and Solutions, W. H. Freeman, San Francisco, 1972.

Emmel, T. C., An Introduction to Ecology and Population Biology, W. W. Norton, New York, 1972.

Jeffers, J. N. R. (editor), Mathematical Models in Ecology, Proceedings of the 12th Symposium of the British Ecological Society, Blackwell Scientific Publication, Oxford, 1972.

MacArthur, R. H., Geographical Ecology, Patterns in the Distribution of Species, Harper and Row, New York, 1972.

Patten, B. C., Systems Analysis and Simulation in Ecology, Vol. 2, Academic Press, New York, 1972.

Watt, K. E. F., Principles of Environmental Science, McGraw-Hill, New York, 1973.

COMPUTERS IN BIOLOGY, MEDICINE

Ledley, R. S., Use of Computers in Biology and Medicine, McGraw-Hill, New York, 1965.

Stacy, R. W., and Waxman, B. D., Computers in Biomedical Research, Vol. 1, Academic Press, New York, 1965.

Stacy, R. W., and Waxman, B. D., Computers in Biomedical Research, Vol. 2, Academic Press, New York, 1965.

Sterling, T. D., and Pollack, S. V., Computers and the Life Sciences, Columbia University Press, New York, 1965.

Li, J. C. R., Numerical Analysis, Edwards Brothers, Ann Arbor, 1966.

Röpke, H., und Riemann, J., Analogcomputer in Chemie und Biologie, Eine Einführung, Springer-Verlag, Berlin, 1969.

Fraser, A., and Burnell, D., Computer Models in Genetics, McGraw-Hill, New York, 1970.

Davies, R. G., Computer Programming in Quantitative Biology, Academic Press, New York, 1971.

Epidemic Simulation for Students in Medicine, Center for Environmental Quality Management, 302 Hollister Hall, Cornell University, Ithaca, N.Y., 1972.

The above bibliography includes some items from biophysics, cybernetics, system theory and medicine. Also some books could have been placed under several headings and rightly, or wrongly, the decision was made here not to repeat any publication under a different heading. No reference has been made to system and parameter identification or to such areas as bio-medical engineering. It is difficult to know where to stop listing material in a bibliography of the present type in an area that appears to have diverse ramifications. Some references have been given to work in quantitative biology, zoology and ecology; however, there are many more publications in these areas alone. In connection with these areas a much more extensive and extremely comprehensive bibliography is under compilation by Prof. V. Schultz, Department of Zoology, Washington State University, Pullman, Washington, 99163, and I understand that all of his references will be available on computer tape at the end of 1973.